中国轻工业"十三五"规划教材
普通高等教育茶学专业教材

茶学专业英语

张正竹　李大祥　主编
[美] Chi-Tang Ho　主审

中国轻工业出版社

图书在版编目（CIP）数据

茶学专业英语/张正竹，李大祥主编.—北京：中国轻工业出版社，2021.6
ISBN 978-7-5184-2267-8

Ⅰ.①茶… Ⅱ.①张…②李… Ⅲ.①茶业—英语—教材 Ⅳ.①TS272

中国版本图书馆CIP数据核字（2020）第039028号

责任编辑：贾　磊　　责任终审：张乃东　　封面设计：锋尚设计
版式设计：砚祥志远　　责任校对：吴大鹏　　责任监印：张　可

出版发行：中国轻工业出版社（北京东长安街6号，邮编：100740）
印　　刷：三河市国英印务有限公司
经　　销：各地新华书店
版　　次：2021年6月第1版第2次印刷
开　　本：787×1092　1/16　印张：15.25
字　　数：340千字
书　　号：ISBN 978-7-5184-2267-8　定价：45.00元
邮购电话：010-65241695
发行电话：010-85119835　传真：85113293
网　　址：http：//www.chlip.com.cn
Email：club@chlip.com.cn
如发现图书残缺请与我社邮购联系调换
210602J1C102ZBW

本书编写人员

主　编

　　张正竹（安徽农业大学）

　　李大祥（安徽农业大学）

副主编

　　林金科（福建农林大学）

　　邓威威（安徽农业大学）

　　张　盛（湖南农业大学）

参　编

　　曹藩荣（华南农业大学）

　　李远华（武夷学院）

　　陈　萍（浙江大学）

　　李品武（四川农业大学）

　　曾　亮（西南大学）

　　王　郁（华中农业大学）

　　温晓菊（浙江农林大学）

　　蒋文倩（安徽农业大学）

　　宋　丽（安徽农业大学）

　　吴亮宇（福建农林大学）

　　石玉涛（武夷学院）

主　审

　　Chi-Tang Ho（Rutgers, The State University of New Jersey, USA）

序

中国是茶树原产地和第一大产茶国，茶叶深受世界人民的喜爱，也是中华传统文化和东西方文化交流的重要载体。随着经济全球化的发展，国际交流变得日益频繁，茶产业发展和茶文化交流也迎来了新机遇，在这个大背景下，加强茶学专业英语知识的学习显得尤为必要。

当前，我国高等教育进入了新时代，一流学科、一流专业、新农科等的建设正全面展开。教材体系建设是一流专业建设的重要内涵，是建设一流课程的基础，更是奠定学生专业功底的载体。据不完全统计，我国现有涉茶高等院校70多所，本科和专科院校涉茶专业年招生量在6000人左右。如何通过茶学高等教育，培养具有国际化视野的高等人才，服务"一带一路"倡议，是新形势下茶学高等教育者面临的共同任务。

值此之际，《茶学专业英语》教材编撰完成。该教材在编写人员上，是一支具有国际化背景的茶学教科研一线的优秀专家团队；在编写内容上，覆盖了茶的历史、茶文化、茶树栽培育种、茶叶生物化学、茶叶加工及机械、茶叶审评、茶与健康、茶叶市场等全产业链，此外还列有科技论文写作内容；在编写素材上，主要内容均源自英文原版书籍和论文。

该书作为我国第一部正式出版的《茶学专业英语》教材，对于丰富和完善茶学专业教学内容，提高茶学人才的专业英语表达、写作、沟通能力，培养茶学国际化人才具有重要意义。

谨此为序。

教授

教授

2020年3月于安徽农业大学

前言

古人云："学然后知不足，教然后知困。"教学相长，互相促进。因多年《茶学专业英语》课程教学实践和茶学国际交流合作的需要，我们决定编写《茶学专业英语》教材。在编写过程中，通过精选世界茶文化与茶学科技知识，使我们再次感受到了茶文化的博大精深、茶学科技的蓬勃发展以及茶给人们带来的美好生活体验。本教材的教学目标是，培养学生的国际化视野和掌握阅读、写作茶学英语科技论文的基本技能，提高英语综合阅读与写作能力，使学生更好地服务于"一带一路"倡议，推动中国茶更好地走向世界。

随着我国茶学科技的快速发展和国际交流的日益频繁，茶学专业英语知识的学习与应用显得日益迫切。基于普通高等教育茶学专业的培养目标，使学生对茶自然科学有更深刻的理解，对茶艺、茶文化有更高的展示和交流能力，本教材内容涵盖了茶与"一带一路"倡议、茶的历史、茶文化、茶树栽培与育种、茶叶生物化学、茶叶加工与机械、茶叶质量评价、茶的健康功效、茶叶营销与贸易以及茶学科技论文写作共10个单元，以期满足学生对茶学专业英语知识系统学习的需要。本教材所选英文材料的主体来自英文原版文章和书籍，并进行了优化处理。附录部分列出了茶类常用机构网站、审评术语等，以供学生课后拓展学习知识时使用。

本教材由张正竹、李大祥担任主编并负责统稿，由美国新泽西州立罗格斯大学 Chi-Tang Ho（何其傥）教授主审。具体编写分工：第一单元由邓威威、张正竹编写；第二单元由蒋文倩编写；第三单元第一课和第二课由宋丽编写，第三~五课由温晓菊编写，第六课由李大祥编写；第四单元第一课、第二课由李远华、石玉涛编写，第三课、第四课由王郁编写；第五单元第一课、第二课由李大祥编写，第三~五课由林金科、吴亮宇编写；第六单元第一课、第四课由李大祥编写，第二课、第三课、第五课由曹藩荣编写，第六~九课由张盛编写；第七单元第一课、第二课由宋丽编写，第三课、第四课由陈萍编写；第八单元第一课、第四~六课由李大祥编写，第二课、第三课由曾亮编写；第九单元由李品武编写；第十单元由邓威威、张正竹编写；附录由李大祥整理。

本教材在中国轻工业出版社有限公司的组织、协调下，在全国涉茶高校一线教师和专家团队通力合作、共同努力下，在 Chi-Tang Ho 教授的仔细审校下才得以顺利完成，在此对他们表示感谢。同时，还要衷心感谢本教材所摘录的原文作者们，他们为本书的出版提供了高质量的素材。最后，特别感谢安徽农业大学宛晓春教授和夏涛教授为本教材作序。

由于编者水平有限，疏漏、错误之处在所难免，恳请广大读者批评、指正。

张正竹、李大祥

2020年3月

英文目录

UNIT One   About Tea 1
  Lesson 1   General Introduction 1
  Lesson 2   Tea and "Belt and Road Initiative (BRI)" 6

UNIT Two   Tea History 9
  Lesson 1   Chinese Tea History 9
  Lesson 2   The Ancient Tea-Horse Road 13
  Lesson 3   Tea History of Japan and Korea 16
  Lesson 4   Tea Comes to the West 20
  Lesson 5   Tea History of India, Sri Lanka and Kenya 23

UNIT Three   Tea Culture 27
  Lesson 1   *Tea Classic* (*Cha Jing*) 27
  Lesson 2   The Chinese Art of Tea 31
  Lesson 3   The Japanese Way of Tea (chanoyu) 38
  Lesson 4   Afternoon Tea 42
  Lesson 5   Indian Tea Culture 46
  Lesson 6   Russian Tea Culture 49

UNIT Four   Tea Cultivation and Breeding 55
  Lesson 1   Botanical Classification of Tea 55
  Lesson 2   Tea Cultivation 61
  Lesson 3   Tea Breeding 66

Lesson 4　Good Agricultural Practice Management in Tea Plantation ………… 70

UNIT Five　Tea Biochemistry ……………………………… 75
Lesson 1　Constituents in Fresh Tea Leaves ……………………… 75
Lesson 2　Biosynthesis of Characteristic Secondary Metabolites ………… 80
Lesson 3　Polyphenol Oxidase ……………………………… 86
Lesson 4　Glycosidase ……………………………………… 90
Lesson 5　Quantification of Catechins, Caffeine and Theanine by HPLC ……………………… 93

UNIT Six　Tea Processing and Machinery ……………………… 99
Lesson 1　Tea Classification ………………………………… 99
Lesson 2　Green Tea ………………………………………… 102
Lesson 3　Black Tea ………………………………………… 107
Lesson 4　Oolong Tea (Blue Tea) …………………………… 110
Lesson 5　White Tea ………………………………………… 115
Lesson 6　Dark Tea ………………………………………… 119
Lesson 7　Yellow Tea ……………………………………… 124
Lesson 8　Further Processing ……………………………… 127
Lesson 9　Tea Storage ……………………………………… 134

UNIT Seven　Tea Quality Evaluation ……………………… 139
Lesson 1　Water Quality …………………………………… 139
Lesson 2　Tea Utensils ……………………………………… 142
Lesson 3　Formation of Tea Quality ………………………… 147
Lesson 4　Tea Evaluation and Inspection ………………… 152

UNIT Eight　Tea and Health Effects …… 161
- Lesson 1　Antioxidant and Pro-oxidant Effects …… 161
- Lesson 2　Anticancer Effects …… 168
- Lesson 3　Effects on Metabolic Syndrome and Related Diseases …… 174
- Lesson 4　Effects on Aging …… 178
- Lesson 5　Bioavailability of Tea Components …… 183
- Lesson 6　Tea Safety …… 191

UNIT Nine　Tea Marketing and Trade …… 195
- Lesson 1　World Tea Production and Consumption …… 195
- Lesson 2　Tea Exports and Imports …… 201
- Lesson 3　Tea Market Development and Outlook …… 206

UNIT Ten　Tea Academic Paper Writing …… 213
- 一、科技论文结构 …… 213
- 二、科技论文写作方法 …… 215
- 三、科研诚信 …… 217

Appendix …… 221
- Appendix A：Websites and Associations (Societies) on Tea …… 221
- Appendix B：Tea Sensory Terminology …… 223
- Appendix C：Introduction to State Key Laboratory of Tea Plant Biology and Utilization …… 227

中文目录

第一单元　绪论 1
第一课　概述 1
第二课　茶与"一带一路"倡议 6

第二单元　茶的历史 9
第一课　中国茶史 9
第二课　茶马古道 13
第三课　日韩茶史 16
第四课　茶在西方传播 20
第五课　印度、斯里兰卡和肯尼亚茶史 23

第三单元　茶文化 27
第一课　《茶经》 27
第二课　中国茶艺 31
第三课　日本茶道 38
第四课　（英式）下午茶 42
第五课　印度茶文化 46
第六课　俄罗斯茶文化 49

第四单元　茶树栽培与育种 55
第一课　茶树植物学分类 55
第二课　茶树栽培 61
第三课　茶树育种 66
第四课　茶园良好农业规范管理 70

第五单元　茶叶生物化学 ·· 75

第一课　茶鲜叶中的化学成分 ·· 75
第二课　特征性次级代谢产物的合成 ·· 80
第三课　多酚氧化酶 ·· 86
第四课　糖苷酶 ·· 90
第五课　儿茶素、咖啡碱和茶氨酸的高效液相色谱法定量 ··························· 93

第六单元　茶叶加工与机械 ·· 99

第一课　茶叶分类 ··· 99
第二课　绿茶 ·· 102
第三课　红茶 ·· 107
第四课　乌龙茶（青茶） ··· 110
第五课　白茶 ·· 115
第六课　黑茶 ·· 119
第七课　黄茶 ·· 124
第八课　再加工茶 ·· 127
第九课　茶叶贮藏 ·· 134

第七单元　茶叶质量评价 ·· 139

第一课　水质 ·· 139
第二课　茶具 ·· 142
第三课　茶叶品质的形成 ··· 147
第四课　茶叶审评与检验 ··· 152

第八单元　茶的健康功效 ·· 161

第一课　抗氧化与促氧化作用 ··· 161
第二课　抗癌作用 ·· 168
第三课　对代谢综合征及相关疾病的作用 ··· 174

第四课　对衰老的影响 ……………………………………… 178
　　第五课　茶叶成分的生物利用度 ……………………………… 183
　　第六课　茶叶安全 ……………………………………………… 191

第九单元　茶叶市场营销与贸易 …………………………………… 195
　　第一课　世界茶叶产量与消费量 ……………………………… 195
　　第二课　茶叶进出口 …………………………………………… 201
　　第三课　茶叶市场未来展望 …………………………………… 206

第十单元　茶叶科技论文写作 ……………………………………… 213
　　一、科技论文结构 ……………………………………………… 213
　　二、科技论文写作方法 ………………………………………… 215
　　三、科研诚信 …………………………………………………… 217

附录 …………………………………………………………………… 221
　　附录A：茶叶协（学）会网址 ………………………………… 221
　　附录B：茶叶感官审评术语 …………………………………… 223
　　附录C：茶树生物学与资源利用国家重点实验室简介 ……… 227

UNIT One About Tea

Lesson 1 General Introduction

Tea originated in China, and people all over the world have taken to drinking it. You might be surprised to realize that (after water) tea is the favorite drink for people from all walks of life, and in all countries, who enjoy the taste, the possibility of health benefits, the wide variety of taste options, and the social interaction that is so closely associated with it. In fact the tea consumed around the world equals all other drinks (coffee, chocolate, soft drinks and alcohol) combined.

The tea industry in China is the largest in the world, and has been dominating the global tea industry for centuries, since China introduced tea to the world. Within China, tea is also important in the economy, as more than 80 million people work in the tea industry as farmers, workers, or sales people. China itself is also the world's largest tea market, or consumer of tea, with over 2 million tons of tea consumed in 2019, averaging 1.5 kg of tea per person per year. However, in terms of how much tea is drunk per person, China is not at the top of the list, with people in Turkey in first place consuming 3.2 kg each person per year, the Irish in second place consuming 2.2 kg each person per year, and the British in third place consuming 1.9 kg each person per year.

While the popularity of tea itself has not declined, people's habits and preferences have been changing. Health conscious people are becoming more aware of the health benefits of particular teas. Also loose-leaf teas are beginning to take root in places where tea bags were previously the most popular form of the beverage. There is more interest in specialty teas, and people are willing to pay more for a quality tea.

1. Biological characteristics

The tea plant [*Camellia sinensis* (L.) O. Kuntze] is an economic woody crop and a perennial plant that is cultivated worldwide. The tea plant is an evergreen shrub that develops fragrant white, five-petaled flowers. Tea plant is traditional propagated through either seeds or stem cutting, with a life span of more than 100 years. Two main varieties are cultivated: *C. sinensis sinensis*, a Chinese plant with small leaves, and *C. sinensis assamica*, an Indian plant with large leaves. Hybrids of these two varieties are also cultivated. In the wild, the tea tree may grow from 5 to 15 m, and sometimes even to 30 m. The wild distribution is in the foothills of the Himalayas, stretching from northeast India to southwest China. Cultivated tea shrubs are usually trimmed to below 2 m to stimulate the growth of leaves and to ease plucking. Tea grows wild in subtropical monsoon climates with wet and hot summers and relatively cold and dry winters. Today, it is cultivated in tropical and subtropical regions. In the tropical regions, the best conditions for tea are at higher altitudes. In most tea growing countries, the existing tea populations were established from seedling with plants of varied mixture with a high proportion of low yielding types. The nursery plants propagated by tea cuttings are true to type but their survival percentage both in the nursery as well as in the field after planting is less as compared to seedlings. It is reported that the period around September/October is the most suitable time for raising tea cutting under plastic sheet in the tea growing areas.

2. Secondary metabolites in tea plants

Secondary metabolism (also called specialized metabolism) is a term for pathways and small molecule products of metabolism that are not absolutely required for the survival of the organism. Tea (*Camellia sinensis*) plants produce unique secondary products, such as caffeine, catechins and theanine. The metabolisms of theanine, caffeine, and catechins are defined as the specific secondary metabolisms in tea.

3. Tea processing and classification

Tea is produced from leaves and leaf buds of the tea plant, *C. sinensis*. All tea varieties, such as green, oolong and black tea, are harvested from this species, they differ in processing. Traditionally, teas are classified into the following six categories based on their respective processing techniques: green tea (enzyme inactivation), black tea (fermentation), white tea (withering), oolong tea (fine manipulation), yellow tea (heaping for yellowing), and dark tea (piling) (Figure 1.1). These processing techniques were developed over a span of thousands of years in different parts of China. The traditional way for identifying tea categories depends primarily on sensory evaluation; teas are classified based on color, aroma, appearance, and taste by trained tea specialists. This is a subjective approach because the results are easily influenced by the environment or taster's experiences. Non-sensory classification may be possible through the use of various instrumental techniques. Many studies have been conducted on the discrimination of tea types; for example, spectroscopy and chromatography have been applied extensively for tea classification in recent years.

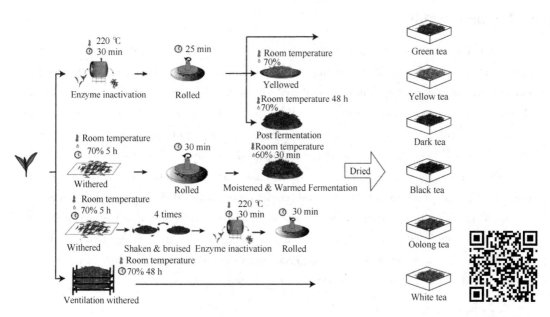

Figure 1.1　Flow diagram depicting the manufacture processes used to produce the six tea types

4. Evaluation and tea culture

　　As teas have traditionally been assessed organoleptically, i.e. with special reference to smell and taste, tea tasting has considered more of an art than a science. To maintain a fresh palate, tea tasters should refrain from eating strong foods (such as curries), drinking alcohol or smoking prior to a tasting session. A discerning palate is needed to distinguish plain teas from quality ones. Master tasters, some of whom have familiarized themselves with the characteristics of up to 1500 individual teas, are capable of assessing 50 teas an hour. Training to become a taster requires at least a five-year apprenticeship, but in reality, the learning process never really ceases. Essentially, judging a tea focuses on cup quality and the leaf.

　　The tea culture includes poems, pictures, and articles about tea and the art of making and drinking tea. It also includes some customs about tea. There are many customs about tea. According to the Travel China Guide on the web, it is customary that the host of the home being visited fill the tea cup to 7/10 of its capacity. It is said that the other 3/10 will be filled with affection and friendship. This tea cup should be emptied in three swallows. Serving tea is more than a matter of politeness; it is a symbol of togetherness, a way to respect the company at your home, and to share something enjoyable. Tea is offered as soon as the guest enters the home and, in some cultures, to not accept a cup of tea or to at least take a sip of the tea is considered rude.

5. Tea and health

Tea is one of the most popular beverages in the world because of its attractive aroma, taste, and healthy effects. Tea contains various beneficial constituents, and several *in vitro* and *in vivo* studies have demonstrated that constituents of tea exhibit biological and pharmacological properties. These constituents have been reported to have antioxidative activity, antimutagenic effects, anticarcinogenic effects, and antiallergic effects. Can drinking several cups of green tea a day keep the doctor away? This certainly seems so, given the popularity of this practice in East Asian culture and the increased interest in green tea in the Western world. Several epidemiological studies have shown beneficial effects of green tea in cancer, cardiovascular, and neurological diseases. The health benefits associated with green tea consumption have also been corroborated in animal studies of cancer chemoprevention, hypercholesterolemia, arteriosclerosis, Parkinson's disease, Alzheimer's disease, and other aging related disorders.

词汇表

Camellia sinensis (L.) O. Kuntze 茶树的拉丁学名

perennial [pəˈrenɪəl] *adj.* 多年生的；常年的；四季不断的；常在的；反复；*n.* 多年生植物

evergreen [ˈevəɡriːn] *n.* 常绿树；常绿植物；*adj.* [植] 常绿的；永葆青春的

caffeine [ˈkæfiːn] *n.* [有化] [药] 咖啡碱，旧译"咖啡因"

catechin [ˈkætɪtʃɪn] *n.* 儿茶素；catechins 表示"儿茶素类"

theanine [ˈθiːənin] *n.* 茶氨酸

secondary metabolism 次生代谢，是相对于初级代谢(primary metabolism)而言。初级代谢主要指能量代谢，氨基酸、蛋白质、核酸、脂肪等的代谢；而次级代谢是指生物合成生命非必需物质并储存次生代谢产物的过程，次生代谢反应仅在特定的物种、器官或组织中于一定的环境和时间条件下才进行。茶树中的咖啡碱、茶氨酸和儿茶素代谢都是次生代谢

fermentation [fɜːmenˈteɪʃ(ə)n] *n.* (食品)发酵；茶叶加工术语中实指以多酚物质为主的氧化

organoleptically [ˌɔːɡ(ə)nə(ʊ)ˈleptɪkəlɪ] *adv.* 感官地；用感官觉察地

palate [ˈpælət] *n.* 味觉

master taster 评尝师

serving tea 奉茶

antioxidative [ˌantəˈɒksədənt] *adj.* 抗氧化的

antimutagenic [æntɪmjuːtɪdˈʒenɪk] *adj.* 抗突变的，抗诱变剂的

anticarcinogenic [æntiˌkɑːsɪnədˈʒenɪk] *adj.* 抑制癌发生的，防癌的，抗癌的

antiallergic [ænʃˈlɜːdʒɪk] *adj. & n.* 抗过敏的

epidemiological [ˌepɪˌdiːmɪəˈlɒdʒɪkl] *adj.* 流行病学

cardiovascular [ˌkɑːdɪəʊˈvæskjələ(r)] *adj.* 心血管的

neurological disease [ˌnjʊərəˈlɒdʒɪkl] 神经(系统)疾病

chemoprevention [kiːməʊpriːˈvenʃn] [医] 化学预防(癌症)，化学预防

hypercholesterolemia [haɪpə(ː)kəlestərəʊˈliːmjə]	*n.* 血胆固醇过多症
arteriosclerosis [ɑːˌtɪərɪəʊskləˈrəʊsɪs]	*n.* 动脉硬化
Parkinson's disease [ˈpɑːkɪnsən]	帕金森病
Alzheimer's disease	阿尔茨海默病
aging related disorder	衰老相关疾病

思考题

1. 将下面的句子译成英文。

（1）茶树是世界上最重要的经济作物之一，是多年生的木本植物，其繁殖方式有种子繁殖或扦插繁殖。茶树的寿命会长达100余年。在一些衰老茶园，可以通过修剪来提高茶叶产量、刺激茶树生长。

（2）茶树中最重要的次级代谢物有茶氨酸、咖啡碱和儿茶素。

（3）根据不同的发酵程度和加工方式，茶叶可分为绿茶、红茶、黑茶、黄茶、乌龙茶、白茶六大类。其中，绿茶是未发酵茶。

2. 将下面的句子译成中文。

（1）As teas have traditionally been assessed organoleptically, i.e. with special reference to smell and taste, tea tasting has come to be considered more of an art than a science.

（2）Yellow teas processed similarly to green tea, but with a slower drying phase called yellowing, which is unique to yellow teas. Dark tea is a relatively unknown form of tea in the west, but is a vital part of the diet of the Tibetan people.

（3）Tea (*Camellia sinensis*) plants produce unique secondary products, such as caffeine, catechins and theanine. The metabolisms of theanine, caffeine, and catechins are defined as the specific secondary metabolisms in tea.

参考文献

[1] WICKHAM R. Chinese tea in trade and the economy [EB/OL]. https://www.chinaeducationaltours.com/guide/culture-chinese-tea-in-trade.htm.

[2] WANG Y, KAN Z, THOMPSON H J, et al. The impact of six typical processing methods on the chemical composition of tea leaves using a single *Camellia sinensis* cultivar, Longjing 43 [J]. Journal of Agricultural and Food Chemistry, 2019, 67: 5423-5436.

[3] SOHAIL A, HAMID F S, WAHEED A, *et al*. Efficacy of different botanical materials against APHID Toxoptera aurantii on tea (*Camellia sinensis* L.) cuttings under high shade nursery [J]. Journal of Materials and Environmental Science, 2012, 3(6): 1065-1070.

Lesson 2 Tea and "Belt and Road Initiative (BRI)"

The Belt and Road Initiative (BRI), also known as the One Belt One Road (OBOR) or the Silk Road Economic Belt and the 21st-century Maritime Silk Road, is a development strategy adopted by the Chinese government involving trade and infrastructure network in more than 60 countries and regions, with a population of 4.4 billion. The president Xi Jinping, originally announced the strategy during official visits to Indonesia and Kazakhstan in 2013. "Belt" refers to the overland routes for road and rail transportation, called "The Silk Road Economic Belt"; whereas "road" refers to the sea routes, or the 21st Century Maritime Silk Road. Until 2016 the initiative was officially known in English as the One Belt and One Road initiative (BRI). The goal of the initiative is "a bid to enhance regional connectivity and embrace a brighter future".

BRI is a major national strategy in current China, while tea used to be the most important commodity on traditional "Belt and Road". Nations along the BRI route produce more than 80% of the world's supply of tea. Tea is a symbol of Chinese culture and hospitality, and plays an extremely important strategic role in the country's foreign trade and cultural exchanges. The connotation of the Chinese character of tea (茶), which is formed with Chinese character people (人) between grass (艹) and trees (木). As long as people can shelve differences, promote harmony and adhere to the principle of harmony in diversity, people can maximize their common interests. When President Xi delivered a speech at the College of Europe in Belgium in 2014, he used the comparison of tea and beer to talk about China-Europe relations. "The Chinese people are fond of tea and the Belgians love beer. To me, the moderate tea drinker and the passionate beer lover represent two ways of understanding life and knowing the world, and I find them equally rewarding. When good friends get together, they may want to drink to their heart's content to show their friendship. They may also choose to sit down quietly and drink tea while chatting about their lives. In China, we value the idea of preserving 'harmony without uniformity', and here in the EU people stress the need to be 'united in diversity'. Let us work together for all flowers of human civilizations to blossom together", said Xi.

Tea ranks among China's most significant contributions to the West. The BRI is not just a rejuvenation of the ancient silk road, but also a comeback of the ancient tea road. Many foreign countries nowadays also come for advanced techniques and cooperation rather than raw tea. Experts also say some tea makers are attempting to deal with more Belt and Road importers to bolster demand for this Chinese specialty. Some have found potential growth areas along the routes, despite the overall flat market share of Chinese tea globally. For Chinese tea companies, attempts to boost trade and increase overseas demand also include promoting tea culture under the Belt and Road Initiative. As tea trade continues on the silk road, China has found a new position in the global supply chain. And its tea makers continue to seize new opportunities in increasing international contacts.

词汇表

Belt and Road Initiative (BRI) "一带一路"倡议，是"丝绸之路经济带"和"21世纪海上丝绸之路"的简称，2013年9月和10月由中国国家主席习近平分别提出建设"新丝绸之路经济带"和"21世纪海上丝绸之路"的合作倡议

UNIT Two   Tea History

Lesson 1   Chinese Tea History

The story of tea is the story of humankind in a nutshell, or perhaps a teacup. It includes the best and the worst of who we are and what we do. Throughout its long history, tea has been used as medicine, as an aid to meditation, as currency, as bribes, and as a means of controlling rebellions. It has been the instigation for wars and global conflicts. It has also been the reason for parties, for family gatherings, and for high-society occasions. In short, tea has touched and changed our lives as no other beverage has.

1. Origin of tea

As the only continuous civilization in the past 5000 years, the length of history on most objects in China would be measured by centuries even millenniums. History of tea is no exception of this realm, being the "birthplace" of tea, Chinese people inherited with a marvelous and prosperous tea industry at present. The most popular and best-known legend about the origin of tea dates to about 3000 BC, during the time of the mythical Chinese emperor Shen Nong (divine farmer), who is said to be the first ever to taste tea. It says Shen Nong, the fictional emperor in prehistoric legends who taught Chinese people to cultivate crops and domesticate livestock, found tea as antidote to poisonous plants accidently in 2737 BC.

Despite the fact that there were no testimonies survive to describe how and why people first consumed tea, it is most likely that the tea tree comes from a place far from the heartland of Chinese culture, remote as mountainous area between Sichuan and Yunnan province, southwest part of China. It has been proved by multiple ways; evidences found in morphology, genetics studies etc.

In the nineteenth century, other theories of tea origin had widespread around the world. Based on the discovery of wild arbor tea tree in Assam area, some botanists and tea sellers distilled, even cooked up new theories of tea origin and this time the credibility of "Shen Nong and Tea" story was deeply questioned.

Take "India-origin theory" for example, ancient Chinese texts mention tea by different name dozens of times, records extend from 59 BC to present days. The first textual reference about using tea in people's life that all Chinese scholars agree to, is recorded by Wang Bao (王褒) in 59 BC, during the ruling of Xuan Emperor Liu Xun (汉宣帝刘询), the Seventh ruler of western Han Dynasty (202 BC-8 AD). In 2016, an unambiguous proof of tea is found in Han Yangling Museum (汉阳陵博物馆), in Xi'an city Shaanxi province; bundles of decayed tea leaves in one of accompanying funeral pit of Jing Emperor Liu Qi (汉景帝刘启) who was the fourth emperor of Han Dynasty and ruled the empire during 157 to 141 BC.

The plant residues unearthed at Han Yangling Mausoleum are the earliest physical evidence of tea in the world, and it might have been buried as royal cuisine ingredient which would be severed to the late emperor in spiritual world, same as the way when he was alive. This archaeological discovery of 2-millennia-old tea samples corroborated textual references of tea using, and date back tea history of China to the second century BC.

Within this continuum of 2000-year-long time span, Chinese people have written about all aspects of tea, emerged with phenomenal "tea book" such as *Tea Classic* since the eighth century, followed with other "tea books" written by ancient scholars in next thirteen centuries. By now, a unique scientific system of tea is conducting in China, research fields stretch in both natural and social science frontiers.

In the meantime, scholars scrutinizing early Indian texts have not found a single reference to tea. In fact, there is no Sanskrit word for tea, and there are no traces of any ancient tea cultivation in India. By the nineteenth century, it is the British under the aegis of the East India Company transformed India into a great tea producer in the nineteenth century with over a million tea plants imported from China. Later, Indian tea plants were transplanted in Indonesia and in the African possessions of the British, but never was a single tea plant imported into China.

An Indian tea origin cannot be supported by any historical fact and legends. It cannot just be coincidence that Indian tea origin theories began to circulate intentionally, at just the time when the British were most anxious to sell the Empire Tea grown in their colonial possessions. Crowning India the "Birthplace of Tea" naturally added prestige to Indian teas and provided a very good selling point.

2. Tea processing history

With multiple scientific researches and results, it is now believed that peoples of mountainous regions far to the southwest of China used tea first, gradually introducing it to areas farther north. By the fourth century, tea was a part of Chinese daily life. People did not drink it for pleasure, however, but continued to use it for its value as a medicine. Lacking a "spoonful of sugar" to help the medicine go down, they tried masking the bitterness with all kinds of additives, including onions, ginger, salt and orange.

Chinese originally regarded tea as an exotic drink, associating it with mysterious subtropical lands far to the south. By the third century, farmers were growing tea in China's far south and west, in what are now Yunnan, Guizhou and Sichuan. Producers and consumers had refined every aspect of tea, from production to drinking techniques. By the fourth century, tea began to infiltrate the lives of ordinary people throughout the south. A robust market led farmers to increase production and also create new hybrids suitable to a wider range of climates. And as growing and processing became increasingly sophisticated, a higher quality and more varied product emerged.

During the time of the Northern Wei Dynasty (386–535 CE), tea leaves were at least primitively processed, and, presumably, the taste improved. A dictionary of this period states that in the district between the provinces of Hubei and Sichuan, tea leaves were harvested, made into cakes, and roasted until hard and reddish in color. The cakes were then pounded into small pieces and placed in a chinaware pot.

Till Tang Dynasty (618–906 AD), tea cultivation was streamlined to a very efficient system permitting each cluster of tea seedlings to yield eight ounces of tea at maturity. Increased tea production became Tang state policy, and for the first time large government plantations appeared.

In Tang Dynasty, when the fresh leaves were brought in they were first steamed and then crushed to form a paste. The tea paste was poured into molds and compressed to form cakes or brick that were perforated, strung together, and baked until dry.

Modern loose tea had not been invented by the Song and Yuan Dynasties (960–1368 AD) and traditional tea manufacture still prevailed. As before, all fresh tea leaves were steamed and then molded into cakes. Song cake tea had a characteristic Long–Feng (龙凤) pattern and addition of fragrance such as ambergris powder in the royal tribute. Most cakes weighed slightly over one pound (≈ 500 g).

The use of loose–leaf tea in the Ming Dynasty (1368–1644 AD) and the consequent development of the teapot, or the massive expansion of literary works on tea during the Ming and Qing Dynasties (1644–1911 AD), were refinements of earlier practices and startling new developments.

The innovation of Chaoqing Tea, or "pan-fried green tea", permitted the manufacture of fine green loose tea that was fragrant and delicate; and if tea was allowed to oxidize naturally until the leaves turned a reddish copper color and then the leaves were spot heated, this controlled factory oxidation resulted in black tea, which the Chinese call "Hong Cha (红茶)". It is the processing tea leaves undergo that determines their final color, shape, aroma and flavor. One of the great tea fallacies holds that green and black teas come from different varieties of tea plants.

When the Ming developed black tea this did not mean that new tea trees had been discovered, because all tea leaves are green on the branch. Both green and black teas can be made from the leaves of the same tree because it is the processing that determines the tea leaves' final aspect. Based on this finding, Chinese tea-men innovated yellow tea, oolong tea, white tea and dark tea in Qing Dynasty, and still being improving ever since.

词汇表

morphology [mɔr'fɑlədʒi] *n.* 形态学(尤指动植物形态学或词语形态学)
genetics [dʒə'netɪks] *n.* 遗传学
paleobotany [ˌpeɪlɪr'bɒtəni] *n.* 古植物学
antidote ['æntidəʊt] *n.* 解药,解毒剂
mausoleum [ˌmɔːsə'liːəm] *n.* 陵墓
Sanskrit ['sænskrɪt] *n.* 梵语;梵文;印度梵文
aegis ['iːdʒɪs] *n.* 庇护,保护;领导
controversial [ˌkɑntrə'vɜrʃl] *adj.* 有争议的,引起争议的,被争论的;好争论的
exotic [ɪɡ'zɒtɪk] *adj.* 异国的;外来的;异乎寻常的,奇异的;吸引人的;*n.* 舶来品
streamline ['strimˌlaɪn] *vt.* 把……做成流线型;使现代化;使简单化;*n.* 流线;流线型
ambergris ['æmbəɡriːs] *n.* 龙涎香(一种脂肪物质,用以制香料)
fallacy ['fæləsi] *n.* 谬误,谬见;谬论;错误;复数: fallacies

思考题

1. 将下面的句子译成英文。

(1) 中国是世界茶叶原产地,而且外销历史悠久,在中国古代贸易中地位突出。自十八世纪始,茶叶出口价值超过传统出口商品的丝绸、瓷器,成为出口创汇的主要货物。

(2) 到了明朝,不再提倡制作饼茶,中国茶叶进入散茶兴盛的时代。在这个历史背景下,茶叶尤其是炒青绿茶发展迅猛。茶叶种植面积扩大,茶叶产量与宋朝、元朝相比提高较快,创制出一些有影响力的茶叶品类。

(3) 随着红茶的出现,人们发现:同一种树叶,竟然可以通过不同的制作方法,被制作成不同的茶产品。除红茶、绿茶之外,又相继创制出黄茶、白茶、乌龙茶和黑茶。

2. 将下面的句子译成中文。

(1) Despite the fact that there were no testimonies survive to describe how and why people first consumed tea, it is most likely that the tea tree comes from a place far from the heartland of Chinese culture, remote as mountainous area between Sichuan and Yunnan province, southwest part of China. It has been proved by multiple ways; evidences found in morphology, genetics studies.

(2) In 2016, an unambiguous proof of tea is found in Han Yangling Museum, in Xi'an

city Shaanxi province; bundles of decayed tea leaves in one of accompanying funeral pit of Jing Emperor Liu Qi who was the fourth emperor of Han Dynasty and ruled the empire during 157 to 141 BC.

(3) In Tang Dynasty, when the fresh leaves were brought in they were first steamed and then crushed to form a paste. The tea paste was poured into molds and compressed to form cakes that were perforated, strung together, and baked until dry.

参考文献

[1] HINSCH B. The rise of tea culture in China: The invention of the Individual [M]. Lanham, MD: Rowman & Littlefield, 2016.

[2] YU F L. Discussion on the originating place and the originating center of tea plants [J]. J Tea Sci, 1986, 6(1): 1-8.

[3] LU H, ZHANG J, YANG Y, et al. Earliest tea as evidence for one branch of the Silk Road across the Tibetan Plateau [J]. Scientific reports, 2016, 6: 18955.

[4] SABERI H. Tea: a global history [M]. London: Reaktion Books, 2010.

[5] BENN J A. Tea in China: A religious and cultural history [M]. Hong Kong: Hong Kong University Press, 2015.

[6] SIGLEY G. The Ancient Tea-Horse Road and the politics of cultural heritage in Southwest China: regional identity in the context of a rising China [M] // Cultural heritage politics in China. Springer, 2013: 235-246.

[7] GAYLARD L. The tea book [M]. London: DK Publishing, 2015.

Lesson 2　The Ancient Tea-Horse Road

1. Origin of "the Ancient Tea-Horse Road"

The unambiguous textual reference to the consumption of tea as a beverage can be dated to 59 BC during the Western Han Dynasty. However, its widespread popularity amongst both northern Chinese and people to the west such as Uighurs is generally attributed to the Tang Dynasty. Previously the oldest physical evidence of tea was from China's Northern Song Dynasty (960-1127 AD). It has long been hypothesized that tea, silks and porcelain were key commodities exported from the ancient Chinese capital, Chang'an (长安), to central Asia and beyond by caravans following several transport routes constituting the network commonly referred to as the Silk Road, in use by the second century BC. There are a few records of tea having been carried along the Silk Road into Tibet, central Asia or southern Asia, however, transportation of tea had only been officially recorded until the Tang Dynasty (618-907 AD).

The tea-horse trade mode by then was considered as the beginning of "the Ancient Tea-Horse Road" (茶马古道, referred as ATHR).

The first clients for "export tea" were the people living outside the Great Wall. Bartering tea for horses had begun in the Tang Dynasty and for almost 500 years the monopoly of this trade belonged to the powerful Horse and Tea Commission (*Chamasi*).

2. Evolvement of "the Ancient Tea-Horse Road"

For a long period since the Song Dynasty (960-1279 AD), dynastic Chinese governments controlled the tea trade between Song, Liao (辽国) and Xixia (西夏) government. This policy held up to Ming dynasty and peaked, in 1406 additional horse markets were established on China's northern frontier in order to revive the tea and horse trade, which had fallen into decay. In its best time, Liaodong (辽东) had over 400,000 horses registered. Balance sheet figures for the tea and horse trade (even accounting for fraud) were staggering. Generally, a good horse "cost" 120 *jin* (≈ 60 kg) of tea, while an average horse was worth 50 *jin* of tea. In the peak year 1389 more than 20,000 horses were exchanged for one million *jin* ($\approx 500,000$ kg) of tea.

Though the Chinese did not take dark tea as their favorite tea at the era, considering its worth of export, they could not ignore its burgeoning importance. Vast plantations controlled by the Horse and Tea Commission were laid out in Shaanxi province expressly to provide tea for the horse trade.

For the south of China, tea trade between Ya'an (雅安) and Tibet conducted through Horse and Tea Commission as well. There are thus substantial records of the amount of trade in both Chinese and Tibetan. More recently other tea producing regions and locations with trading routes with Tibet have also begun to assert their connections to the "tea road". There is now a growing body of research emerging from Qinghai and China's Northwest.

Profits went, at least in theory, into the Ming treasury, but malignant graft in the Horse and Tea Commission reached truly scandalous proportions and the government hardly profited at all. Corruption and smuggling became so rife that action finally had to be taken, and after the death of Emperor Yongle (明永乐帝朱棣) his successor abolished the Horse and Tea Commission on September 7th, 1424. However, the smuggling, doctoring, and faking of teas would remain a worrisome plague throughout the Ming Dynasty.

The Ancient Tea-Horse Road (ATHR) was restarted in beginning of Qing Dynasty (1644-1911 AD), reached its zenith during the late Qing and first half of the twentieth century; and gradually formed tea-horse trade network we are more familiar with: trading routes linking the tea producing areas of Yunnan [mainly concentrated in the southeast of the province in Pu'er (普洱) and Xishuang Banna (西双版纳)] and Sichuan (around the area known as Ya'an) with the tea-consuming regions across China, but in particular with Tibet. The network extended into mainland Southeast Asia, Nepal and India, thus also embodying an international dimension.

However, the emergence of the ATHR has not been without controversy in scholarly circles. Scholars outside Yunnan, and in particular in Sichuan, have expressed concern that the "tea road" has been hijacked. The critics argue that while the concept of a "tea road" between Yunnan and Lhasa is valid, it lacks substantial material evidence to match the claims of historical continuity and significance that is attributed to it. They argue that the tea road between Ya'an (雅安) and Tibet has a much stronger claim. The ATHR as a concept covering the exchange of tea and culture through the mountainous trading networks between Southwest districts and Tibet is now expanding.

Scholars describes the ATHR as a "moving culture", referring to the very intrinsic nature of a "road" as an object built for the transfer of people, goods, and ideas. In historical moment, the caravans have been replaced by trains, planes, automobiles and bullet trains, the muleteers by tourists, and the tea by the consumption of nostalgia and the experience of leisure. This indeed is the ATHR of the twenty-first century.

词汇表

unambiguous [ˌʌnæmˈbɪɡjuəs] *adj.* 不含糊的；清楚的；明白的
Uighur 维吾尔；维吾尔族；维吾尔语
burgeon [ˈbɜːdʒən] *v.* 迅速生长，迅速发展
Horse and Tea Commission (宋、明、清时期) 茶马司
Tea-Horse Trade 茶马贸易
rife [raɪf] *adj.* 流行的；普遍的；盛传的
doctoring [ˈdɒktərɪŋ] *n.* 篡改；伪造
Lhasa *n.* 拉萨

思考题

1. 将下面的句子译成英文。

（1）自宋朝始，中原王朝便掌握了与辽国、西夏政权之间的茶马贸易。此项政策在明朝得到承续且不断发展；至1406年，为挽救已经出现颓势的茶马贸易，朝廷又与边境开放了更多的马市。

（2）时至今日，茶马贸易的概念已经从原先仅指西南省份与西藏山区之间茶叶及茶文化的传播交流，扩展至全国范围内的贸易与交流。

2. 将下面的句子译成中文。

（1）It has long been hypothesized that tea, silks and porcelain were key commodities exported from the ancient Chinese capital, Chang'an (长安), to central Asia and beyond by caravans following several transport routes constituting the network commonly referred to as the

Silk Road, in use by the second century BC.

(2) The first clients for "export tea" were the people living outside the Great Wall. Bartering tea for horses had begun in the Tang Dynasty and for almost 500 years the monopoly of this trade belonged to the powerful Horse and Tea Commission.

(3) Though the Chinese did not take dark tea as their favorite tea at that era, considering its worth of export, they could not ignore its burgeoning importance. Vast plantations controlled by the Horse and Tea Commission were laid out in Shaanxi province expressly to provide tea for the horse trade.

参考文献

[1] EVAN J C. Tea in China: The history of China's national drink [M]. New York: Greenwood Press, 1992.

[2] LU H, ZHANG J, YANG Y, et al. Earliest tea as evidence for one branch of the Silk Road across the Tibetan Plateau [J]. Scientific reports, 2016, 6: 18955.

[3] SIGLEY G. The Ancient Tea-Horse Road and the politics of cultural heritage in Southwest China: regional identity in the context of a rising China [M]. // Cultural heritage politics in China. Springer, 2013: 235-246.

[4] MU J. Ethnic Culture along the ancient tea-horse road [M]. Kunming: Nationalities Publishing House of Yunnan, 2003.

Lesson 3 Tea History of Japan and Korea

Because drinking tea soothed the mind but kept one alert and awake, Buddhist monks frequently used it as a tool for meditation. As monks traveled from one country to the next, teaching about Buddhism and meditation, they took tea with them, and so the habit of drinking tea flowed from China throughout Southeast Asia and beyond.

1. Tea history of Japan

The knowledge of tea was probably brought to Japan from China in the late sixth century at the same time as Buddhism. During the Tang dynasty many Japanese priests pursued their studies of Buddhism in China and brought back various customs, including the use of cake tea (called *dancha* in Japan), which was the favored drink of China at that time.

The origin of Japanese complex religious and cultural associations with tea found in Kissa Yojoki (*Drinking Tea for Nourishing Life*《吃茶养生记》) by the Japanese monk Eisai (荣西禅师, 1141-1215 AD). Eisai spent a considerable period of time in China, his firsthand experience with the consumption of tea and other decoctions there informs his work. He offers a unique perspective on religious and cultural aspects of tea in China, including important eyewitness accounts of methods of tea production and consumption in Zhejiang in late Southern Song in the twelfth century. Second, his text contains many elements taken from the continent that are rearranged in a new and distinctive congratulation. Eisai's creative display of knowledge, techniques, concepts, and language from the mainland offers an excellent opportunity to respect on the meanings of tea in China. During the thirteenth century the custom of drinking *matcha* (抹茶) first reached the temples and the upper level of samurai society. It then spread to the common people, who began to drink tea as an enjoyable beverage rather than as a health drink or medicine.

In the late fourteenth century a Zen priest called Shuko (AKA Murata Shuko, 村田珠光, 1423-1502 AD) combined the rituals of preparing and drinking of tea with a spiritual sense of humility and tranquility, thus creating the tea ceremony. He became the first great tea master and it was he who prepared an austere code of rules. Tea-drinking and the rituals connected with it play a very important part in the Japanese way of life. The tea ceremony, cha-no-yu, which means "hot-water-tea" and is also called *chado* (茶道), "The Way of Tea", was originally a Buddhist ritual. Monks drank powdered tea from a communal bowl before an image of Buddha.

In the sixteenth century Zen tea master Sen Rikyu (千利休, 1522-1591 AD) amended the ceremony to the form it has today. The ritual is less elaborate and focuses on harmony, respect, purity and serenity. When asked by one of his disciples what the most important things were in following the "Way of Tea", he proposed the observance of seven rules:

Make a delicious bowl of tea;
Lay out the wood charcoal to heat the water.
Arrange the flowers as they are in the fields.
In summer, evoke coolness; in winter, warmth.
Anticipate the time for everything.
Be prepared for rain.
Show the greatest attention to each of your guests.

The tea ceremony influenced all Japan's fine arts, including garden design, flower arrangement, architecture, calligraphy, painting, lacquer and ceramic arts. It has developed into an elaborate social custom.

It was not until the late sixteenth century that loose leaf tea was imported from China and was known as *Sencha* (煎茶) in Japan. *Sencha* became Japan's most popular everyday tea because it was cheaper and easier to make than the powdered *matcha* tea. Today many kinds of leaf tea are drunk in Japan. Most are green teas but some black teas imported from Ceylon and India are popular and are on the menus of leading hotels and restaurants. There is a Japanese proverb, "*if a man has no tea in him, he is incapable of understanding truth and beauty.*"

2. History of Korea

Like its neighbors, Korea has a rich tea-drinking culture. Green tea (nok cha, Korea) was introduced from Tang Dynasty in China during the reign of Queen Sŏndŏk (善德女王, 632-647 AD) although tea drinking might have been known much earlier. Tea was initially prized for its medicinal properties and was also reserved for special occasions. King Munmu (文武王), who ruled from 661 to 681, ordered tea to be used during ceremonial offerings.

Tea-drinking in Korea is linked with the Panyaro (The Korean Way of Tea) Seon (Zen) of tea and is viewed as a spiritual, religious activity leading to higher levels of inner awakening, if not total enlightenment. Buddhist monks commonly drank tea as an aid to meditation and offered tea to the Buddha three times daily. Temples also served tea to visitors. Due to such demand for tea, villages arose near to temples that cultivated tea and became known as *tach'on*, or "tea villages".

It was not until the reign of the 42nd Silla monarch (新罗君主), King Heungdeok (兴德王, 826-836 AD), that a royal envoy Kim Taeryom returned from a mission in Tang Dynasty and brought seeds of the tea plant. The king ordered the seeds to be planted on the warm slopes of Mount Chiri (智异山), which is still the centre of tea cultivation in Korea.

The ceremonies related to drinking tea at the royal court and elsewhere developed into the custom known as the tea ceremony (tado). Specialized implements for this ceremony were developed, such as a brazier for boiling water, bowls for water and tea, spoons and pots. Types and qualities of tea were also developed, as was a grading system for the taste of water. As in Japan, tea ceremony etiquette is very important in Korea but the harmony of water and tea is even more central to the ritual. Chouiseonsa (草衣禅师, 1786-1866 AD), the famous monk and tea-master, wrote: "In brewing, delicacy, in storing, aridity, in steeping, purity. Delicacy, aridity, and purity are essential to the tea ceremony."

During the time of Ming and Qing Dynasties, loose leaf green tea took over from the old powdered form. After the death of the third Yi monarch (李氏朝鲜第三位君主太宗李芳远, 1367-1422 AD), Buddhist court ceremonies were abruptly replaced by Confucian rites, and wine became the formal drink except in monasteries. The monks, abstainers from alcohol, preserved the custom of drinking tea which through its long association with Buddhism had by then largely become a monastic custom, so that the tea art was thenceforth preserved in an austerely simple form. As the aristocrats showed few signs of abjuring the officially disfavored beverages, the new Confucian-style government levied a heavy tea tax, thereby compelling most monasteries to reduce or destroy their tea crop. Only in the south did some tea plantations survive, so tea drinking declined even in the monasteries, until the partial revival that took place towards the end of the dynasty. Its leader, Chouiseonsa, emphasized the complementarity of tea and meditation.

词汇表

decoction [dɪˈkɒkʃən] *n.* 煎煮，煮出的汁，煎熬的药；煎法；熬出物；汤液
samurai [ˈsæmuraɪ] *n.* (旧时日本的) 武士
AKA (also known as)，也就是
austere [ɒˈstɪə(r)] *adj.* 朴素的；简陋的；严厉的；苦行的；禁欲的
communal bowl 公用碗
Zen *n.* 禅，禅宗(中国佛教宗派)
disciple [dɪˈsaɪpl] *n.* 信徒，追随者；门徒，弟子
harmony, respect, purity and serenity [səˈrenətɪ] 和、敬、清、寂
flower arrangement 插花
"In brewing, delicacy, in storing, aridity, in steeping, purity. Delicacy, aridity, and purity are essential to the tea ceremony." 造时精、藏时燥，泡时洁，精、燥、洁，茶道尽矣
abjure [əbˈdʒʊə(r)] *vt.* 发誓放弃；郑重放弃(意见)

思考题

1. 将下面的句子译成英文。

（1）自公元六世纪始，茶与佛教一起由中国传入日本。当时，日本僧侣前往大唐修佛问道，将知识与大唐的众多习俗文化一同带回日本，包括饮用团茶的习惯。

（2）日本僧人村田珠光于十四世纪晚期，终于将备茶、饮茶的仪式融为一体，并赋予这些仪式以精神层面的意义，首次将人性与空灵注入茶中，创造出一整套仪程，是为茶道。

（3）李氏朝鲜第三位君主，太宗李芳远过世以后，佛教寺院茶礼迅速被儒家典礼取代，酒则成为除寺院以外的场所的正式饮品。

2. 将下面的句子译成中文。

（1）During the time of Ming and Qing Dynasties, loose leaf Green tea took over from the old powdered form. After the death of the third Yi monarch, Buddhist court ceremonies were abruptly replaced by Confucian rites, and wine became the formal drink except in monasteries.

（2）Eisai spent a considerable period of time in China, his firsthand experience with the consumption of tea and other decoctions there informs his work. He offers a unique perspective on religious and cultural aspects of tea in China, including important eyewitness accounts of methods of tea production and consumption in late Southern Song Zhejiang in the twelfth century.

（3）The tea ceremony influenced all Japan's fine arts, including garden design, flower arrangement, architecture, calligraphy, painting, lacquer and ceramic arts. It has developed into an elaborate social custom.

参考文献

[1] SABERI H. Tea: a global history [M]. Reaktion Books, 2010.

[2] BENN J A. Tea in China: A religious and cultural history [M]. Hong Kong: Hong Kong University Press, 2015.

[3] BLOFELD J. The Chinese art of tea [M]. Boston: Shambhala Publications, 1985.

Lesson 4 Tea Comes to the West

In Europe, the sixteenth century dawned with great excitement over sea trade and exploration. The first European port city to experience tea was Amsterdam, during the first few years of the seventeenth century. At first, tea was treated as nothing more than a novelty—though a very expensive one. Tea didn't make it to London for another half-century, but once the Brits found a taste for tea, they were never the same again. The first advertisement for tea appeared in the British weekly magazine the *Mercurius Politicus*（政治快报）, in September 23rd of 1658: "*That Excellent, and by all Physitians approved, China Drink, called by the Chineans, Tcha, by other Nations Tay alias Tee, is sold at the Sultaness-head, a Cophee-house*（苏丹王妃咖啡馆）*in Sweetings Rents by the Royal Exchange, London.*"（注:语言为早期近代英语）The British developed such a mania for tea (fueled by the British East India Company merchants who made vast fortunes selling tea) that it quickly became part of the national culture. Tea the drink and tea the social occasion became a part of British life, for everyone from lords and ladies to the men and women of the working class.

1. Tea history of European countries

Tea is grown in over 60 countries around the world, it began to travel as a trade item as early as the fifth century with some sources indicating Turkish traders bartering for tea on the Mongolian and Tibetan borders. Tsar Alexis of Russia was the recipient of a gift of green tea by courtesy of the Chinese in 1618. Certainly Arab traders had dealt in tea prior to this time, but no Europeans had a hand in tea as a trade item until the Dutch began an active and lucrative trade early in the seventeenth century. Dutch and Portuguese traders were the first to introduce Chinese tea to Europe.

The opening of European markets during the Wanli Period（明万历年间, 1573-1620 AD）when tea reached western Europe after months of sea travel from China. The Portuguese shipped it from the Chinese coastal port of Macao; the Dutch brought it to Europe via Indonesia. From Holland, tea spread relatively quickly throughout Europe.

The Portuguese were familiar with tea in the early sixteenth century and by 1577 they were engaged in trading the commodity with the Chinese on a regular basis, predating the arrival of both the Dutch and English in the Far East. Europe's exposure to tea continued to grow. The curiosity of Queen Elizabeth I of England was sufficiently aroused on hearing the news of the beverage in 1598 as to attempt to acquire some.

The new craze found its way into France around 1635 via the Dutch, but within 50 years interest had waned. Despite gaining acceptance in royal circles, tea never usurped the coffee and wine tradition. France's lasting contribution to tea drinking was the addition of milk to the beverage, thus enhancing its value as a source of extra nourishment. The anchoring in Chinese waters in 1635 of the first East India Company ship, the London, did not lead to the loading of tea by the British for the homeward journey; they were, however, soon to emulate their entrepreneurial Dutch seafaring neighbors who were also importing the highly desirable Yixing teapots. Tea eventually came to Britain around 1646.

With the appearance of tea on the beverage list of Thomas Garway's London coffee shop in Exchange Alley in 1657, the demand for tea had been created. This demand was soon to burgeon into a vibrant tea trade. The association between tea and Britishness is a long established concept which has received considerable scholarly and popular attention. Focusing on the visual and material culture of tea in Britain in the period between 1850 and 1900, it explores how tea was invested with numerous identities and contradictory ideologies of work and rest; luxury and necessity; and of the domestic, industrial and imperial simultaneously. These ideologies were mediated through material culture which shaped the behaviors of the tea table and the reception of foreign products, and through the visual language of tea which it communicated and disseminated ideological meanings. Looking specifically to the second half of the nineteenth century, when the tea trade was at its most complex.

2. Tea and "the Opium War"

The eighteenth century was an exciting time of expansion and discovery for Europeans, while China kept in traditional patterns made the Chinese resistant to an exchange with the West, whether the exchange was of trade goods or ideas. Nevertheless, in 1685, the emperor of the Qing dynasty, decided to open all ports (with stiff duties to safeguard Chinese interests) to Europeans. In 1715 this decision was revoked, however, and all ports except Canton (Guangdong) were closed to foreigners. Trade between China and Europe was restricted to Canton for 160 years.

However, the obsession for tea in England during the nineteenth century had devastating effects half a world away in China and India. As England expanded her imperialistic powers, she became greedier for tea and the profits it engendered. Up until the time of the American Revolutionary War, most of the silver used by England had come from Central and South America. After the war, supplies from Mexico were essentially cut off, and inflation led to a rise in the cost of silver as well. However, the demands for Asian goods especially teas in Britain was on the increase, directors of East India company seized upon the idea of growing poppies and selling opium. They realized that trading opium for tea was more lucrative than buying tea with silver, they quickly developed a huge opium industry in India. The result was hunger and deprivation in India and the Opium War and their tragic toll in China.

3. Boston tea party

As the Dutch spread tea around the world during the seventeenth century, Peter Stuyvesant, Dutch governor in the American colonies, brought the first tea to New Amsterdam in 1647—interestingly enough, ten years before it was introduced to London. Early settlers quickly learned to love their tea. After New Amsterdam was captured by the English in 1674 and renamed New York, the British institutions of the coffeehouse and pleasure garden were brought to the New World. In 1678, William Penn founded Philadelphia. His writings and diaries suggest that tea was his preferred drink.

The popularity of tea took on monumental importance, of course, as a symbol of the American Revolution. The tax on tea that Parliament was imposing in its own country was also applied to Americans, with disastrous results for the British. On December 16th, 1773, a band of angry colonists gathered at Griffin's Wharf in Boston, disguised as native American Indians. They boarded three East Indian Company ships and threw their tea cargoes into Boston Harbor, as a protest against the unfair taxation. These acts and others ultimately led to the Revolutionary War. For a while, drinking tea was seen as unpatriotic, and citizens showed support by switching from tea to coffee or other substitutes. Tea never became the national obsession in America that it is in England—coffee seems to fill that role in the U.S. However, people resumed drinking tea after the war, eventually the United States sent ships to China and began importing tea directly, bypassing the powerful East India Company. Eventually, Americans invented the first teabag and credited with "discovering" iced sweetened tea.

词汇表

Tsar Alexis 沙皇亚历克西斯
lucrative [ˈluːkrətɪv] adj. 获利多的，赚钱的；合算的
usurp [juːˈzɜːp] v. 取代；篡夺；侵占，霸占
Caribbean [ˌkærɪˈbiːən] n. 加勒比海
proletariat [ˌprəʊləˈteərɪət] n. 工人阶级，无产阶级
obsession [əbˈseʃn] n. 着魔，萦绕；使人痴迷的人(或物)

思考题

1. 将下面的句子译成英文。

(1) 1658年9月23日，世界上第一则茶叶广告诞生，在伦敦的《政治快报》刊出：

"为所有医师所认可的极佳的中国饮品。中国人称之茶,而其他国家的人则称之 Tay 或者 Tee。位于伦敦皇家交易所附近的斯维汀斯-润茨街上的苏丹王妃咖啡馆有售。"

(2) 18 世纪茶叶慢慢风行欧洲以后,茶叶进口才成为对华贸易中最主要商品,但需用白银交换。为平衡茶叶贸易造成的巨额逆差,英国人采用被称为"黑色金子"的鸦片来换回白银。

(3) 1773 年英国进一步加强了对北美殖民地的控制,东印度公司垄断了北美殖民地的茶叶销售,并征收印花税。波士顿美国人为了抗议茶叶印花税,在 1773 年 12 月 16 日装扮成印第安人混到这些船只上,把茶叶直接倒进了大海,这就是波士顿倾茶事件。此事件成为美国独立战争的前奏。

2. 将下面的句子译成中文。

(1) The first European port city to experience tea was Amsterdam, during the first few years of the seventeenth century.

(2) Tea first arrived in the west via overland trade into Russia. Tsar Alexis of Russia was the recipient of a gift of green tea by courtesy of the Chinese in 1618.

(3) The opening of European markets during the Wanli Period (1573-1620 AD) when tea reached western Europe after months of sea travel from Chinese. The Portuguese shipped it from the Chinese coastal port of Macao; the Dutch brought it to Europe via Indonesia. From Holland, tea spread relatively quickly throughout Europe.

参考文献

[1] MARTIN L C. A history of tea—The life and times of the World's Favorite Beverage [M]. Vermont: Tuttle Publishing, 2018.

[2] HEISS M L, HEISS R J. The story of tea a cultural history and drinking guide [M]. New York, U.S. Ten Speed Press, 2007.

[3] MARTIN L C. Tea: The drink that changed the world [M]. Vermont, U.S.: Tuttle Publishing, 2007.

Lesson 5　Tea History of India, Sri Lanka and Kenya

1. Tea history of India

There are no doubts that China is the original home of tea. Although wild tea trees may have been growing in Assam from early times, the tea plant was not cultivated there until the British started growing and processing tea on a commercial scale in the 1830s and 1840s.

In 1800, the East India Company was selling about 20 million pounds worth of high-priced Chinese tea in England, where the upper and middle classes had fervently taken to the beverage. For long a consumption good signifying taste and refinement in East Asia, tea, conjoined with abundant sugar supplies from the Caribbean colonies, would soon become the British Empire's drug food of choice, an important and cheap staple for the British working classes. This was made possible through the introduction of commercial tea cultivation into Assam, a region annexed to British-ruled India in the early nineteenth century.

When the East India Company started searching for an alternative source of supply to Chinese tea, Indian tea was first introduced in world market by the British proletariat who developed a habit of strong Indian tea. By 1888 tea from India, mostly grown in Assam, had outstripped China in the world economy, obtaining 57 per cent of the British market. This article traces the interface of science, ideology and economy which integrated Assam into imperial and global commodity networks through its new character as Britain's tea garden.

2. Tea history of Sri Lanka

Sri Lanka, formerly a British colony called Ceylon, is a small and vibrant island nation is famed for the array of high-quality teas that it grows and produces using traditional methods. Originally a coffee-growing nation, Sri Lanka switched to tea production in 1869, when a devastating blight infested the majority of its coffee plantations. By 1890, the export of coffee from Ceylon was less than a tenth of what it had been at its zenith. When coffee could no longer be grown, many of the estates changed to planting tea. As in India, the majority of the plantations were owned by the British. Many of the workers on Ceylon's tea plantations came from India, particularly after droughts in India caused famine, which happened in 1877. During this year, 167,000 Indian people of Tamil descent went to Ceylon to work. The large numbers of immigrants of Tamil descent into Ceylon during the nineteenth century were resented by the native Sinhalese, which set the stage for the devastating ethnic conflicts that exist in Sri Lanka today.

3. Tea history of Kenya

Tea was introduced by the British in 1903, but at first only on a limited scale. Two British companies, Brooke Bond and James Finlay, changed this in the 1920s, when they were able to purchase massive amounts of land very cheaply, although with some controversy. Land in Kenya was made available to British ex-servicemen after World War I, a fact that was naturally greatly resented by the African landowners. Twenty-five thousand acres had been set aside for growing flax, a scheme engineered by fifty-five British ex-offices. Unfortunately, the flax market collapsed, and the entire enterprise fell apart. The land was put on the market for almost nothing and was quickly snatched up by the companies James Finlay (now called African Highlands) and Brooke Bond (now part of Unilever), which were both determined to grow tea there. This part of Kenya proved to be excellent for growing tea. Other British companies began buying and planting land in Kenya as well, until 1976, when the Kenyan government finally was able to stop further expansion of the British tea industry. As of the year 2000, the British companies had planted over fifty thousand acres of tea. Brooke Bond was the biggest owner.

In 1976, the Kenya Tea Development Authority Insurance Agency was founded. This organization, funded by the World Bank, encouraged individual African landowners to grow tea. Although each landowner only planted a small amount, usually only about one acre, so many farmers became involved in the project that small landholders now account for 60 percent of the tea produced in Kenya. June 2000 brought change to this group, as it switched from being a parastatal agency (one wholly or partially owned by the government) to a public company. The name was changed to the Kenya Tea Development Agency (KTDA).

词汇表

East India Company 东印度公司(英属)
upper and middle class 中上层阶级
fervently *adv.* 热烈地；热情地；强烈地
Caribbean [ˌkærɪˈbiːən] *n.* 加勒比海地区
working classe 工人阶级
proletariat [ˌprəʊlɪˈteərɪət] *n.* 无产阶级；工人阶级
blight [blaɪt] *v.* 损害；妨害 *n.* (农作物等的)枯萎病，疫病
zenith [ˈzenɪθ] *n.* 鼎盛时期；顶峰；天顶(太阳或月亮在天空中的最高点)
Tamil descent 泰米尔人的后裔
Sinhalese [ˌsɪnhəˈliːz] *n.* 僧伽罗人(居住在斯里兰卡)；僧伽罗语
ex-servicemen [eksˈsɜrvəsmɛn] *n.* 退役军人；复员军人
flax [flæks] *n.* 亚麻；亚麻纤维
Kenya Tea Development Authority Insurance Agency 肯尼亚茶叶保险局
Kenya Tea Development Authority (KTDA) 肯尼亚茶叶发展局
parastatal agency 半国营机构

思考题

1. 将下面的句子译成英文。

(1) 1869年，斯里兰卡因一场大面积植物枯萎病感染，由原先咖啡种植国家转型成为种茶国。而在茶叶国际贸易市场上，人们仍旧惯用当时的旧称"锡兰茶"来代指斯里兰卡茶叶。

(2) 葡萄牙人先于荷兰与英国人认识茶叶，他们于十六世纪早期来到远东地区并接触茶叶，且在1577年前后，便与中国人就茶叶商品达成日常贸易。这使得欧洲人开始

了解并熟悉茶叶。

（3）肯尼亚于1903年开始在蒙巴萨的LMURU地区首次引种茶叶，到20世纪20年代中期，茶叶开始作为商品生产发展起来，现已成为世界上主要的红茶生产国和贸易国。

2. 将下面的句子译成中文。

（1）By 1888 tea from India, mostly grown in Assam, had outstripped China in the world economy, obtaining 57 per cent of the British market.

（2）Sri Lanka, formerly a British colony called Ceylon, is a small and vibrant island nation is famed for the array of high-quality teas that it grows and produces using traditional methods. Originally a coffee-growing nation, Sri Lanka switched to tea production in 1869, when a devastating blight infested the majority of its coffee plantations.

（3）Although each landowner only planted a small amount, usually only about one acre, so many farmers became involved in the project that small landholders now account for 60 percent of the tea produced in Kenya.

参考文献

［1］MARTIN L C. A history of tea— The life and times of the World's Favorite Beverage ［M］. Vermont, U.S.：Tuttle Publishing, 2018.

［2］HEISS M L, HEISS R J. The story of tea a cultural history and drinking guide ［M］. New York, U.S.：Ten Speed Press, 2007.

［3］MARTIN L C. Tea：The drink that changed the world ［M］. Vermont, U.S.：Tuttle Publishing, 2007.

UNIT Three Tea Culture

Lesson 1 *Tea Classic* (*Cha Jing*)

In the Tang Dynasty (618-907 AD), one of China's golden ages, tea drinking became an art. When we talk about Chinese tea culture today, we shouldn't miss an important figure in Chinese history—Lu Yu (733-804 AD), who was respected as a "Tea Sage" or "the Father of Tea" for his contribution to Chinese tea culture. His major event in tea history was the publication of the first tea book, *Cha Jing*, which is now known as "the *Classic of Tea*" or "*the Tea Classic*" (Figure 3.1).

Born in 733 AD in the Tang Dynasty, Lu Yu was an orphan adopted by a monk in Hubei province. At that time, drinking tea became a nationwide tradition. (Tea drinking originally appeared in Southern China, and until the mid-Tang Dynasty it started to gain favor with Northern Chinese). The widespread distribution of tea can be attributed to the extensive practice of Zen Buddhism in the whole country. Because sleeping and eating were strictly prohibited for Buddhists practicing meditation, they could only drink tea. Many monks were tea connoisseurs at the same time.

The monk who adopted Lu Yu was a tea lover and Lu Yu prepared tea for him from childhood. As the years passed, Lu Yu's skill at preparing tea improved and he developed a great interest in the brew. In his late years, Lu Yu withdrew from the outside world and concentrated on research into tea. The fruits of his research were written down in his masterpiece—*Cha Jing*.

In the book, Lu Yu tried to comprehensively present all known information about Chinese tea culture. It is divided into three sections and ten chapters, including the origin of tea, tea tools, tea picking, tea cooking, tea ceremony and famous tea producing areas. Perhaps of most historical value is the seventh chapter, entitled "Tea events" and records incidents concerning tea over thousands of years, from legendary times to the Tang Dynasty. The ten chapters are as the following:

Part I Origin (一之源), characteristics, name, and qualities of tea.

Part II Tools (二之具), for plucking/picking, steaming, pressing, drying and storing tea leaves and cakes.

Part III Making (三之造), recommended procedures for the production of cake tea.

Part IV Utensils (四之器), for brewing and drinking tea. (The materials and methods for making utensils as well as the size and the functions of each utensil.)

Part V Boiling (五之煮), methods of infusing tea and the water of various places.

Part VI Drinking (六之饮), habits of tea drinking. (the origin of tea drinking, tea spread, and tea drinking customs.)

Part VII History (七之事), stories about tea from ancient times to Tang Dynasty.

Part VIII Growing regions (八之出), which kinds of tea are better in different locations.

Part IX Simpilify (九之略), utensils which may be omitted (nonessential utensils) and under what circumstances.

Part X Pictorial (十之图), how to copy this book on silk scrolls that provide an abbreviated version of the previous nine chapters for everyone to understand.

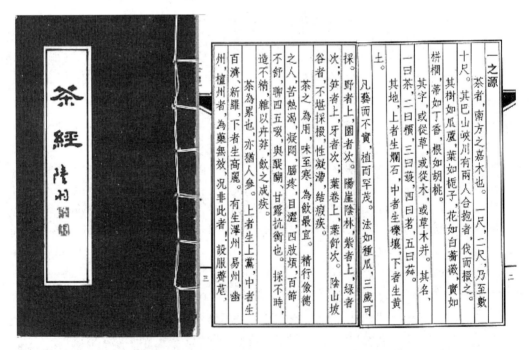

Figure 3.1　Cover and book of *Cha Jing* (*the Classic of Tea*)

This book contributed to turning tea drinking into an art and helped to popularize the art of tea drinking across the country. It played a great role in giving tea cultural significance. The neighboring countries Korea, Japan and Southeast Asia also adopted the tea drinking custom. Lu Yu was the first to suggest that the ritual of preparing and drinking tea represented a code of symbolic harmony and order which reflected the ideals of the cosmos and society.

In his book, *Cha Jing*, Lu Yu lists more than twenty-four implements that are essential at the time for the correct preparation of a cup of tea as followings (Figure 3.2).

- *crushing block* (砧椎)
- *brazier* (风炉)
- *charcoal basket* (炭筥)
- *charcoal mallet* (炭挝)
- *fire chopsticks* (火夹)
- *cauldron* (镇)
- *cauldron stand* (交床)
- *tea tongs* (夹)
- *paper wallet* (纸囊)
- *crushing roller* (碾)
- *sieve box* (罗合)
- *tea holder* (则)
- *water vessel* (水方)
- *water filter bag* (漉水囊)
- *gourd scooper* (瓢)
- *bamboo tongs* (竹夹)
- *salt container* (鹾簋, cuò guǐ)
- *boiled water vessel* (熟盂)
- *bowl* (碗)
- *bowl basket* (畚)
- *brush* (劄)
- *water basin* (涤方)
- *spent tea basin* (滓方)
- *tea cloth* (巾)
- *utensil table* (具列)
- *utensil basket* (都篮)

(1) brazier   (2) cauldron   (3) water filter bag   (4) cauldron stand

(5) salt container   (6) boiled water vessel   (7) water basin   (8) spent tea basin

Figure 3.2   Partial tea implements

These include equipment needed for roasting and grinding the cake tea, a stove for boiling the water, and cups for serving tea. Lu Yu suggested that the brazier for boiling water could be made of brass or iron and shaped like an ancient Ding which was used for sacrifice and formal ceremony. Its three legs represent water, wood and fire, which indicate a proper balance of the five elements in Chinese culture. Besides the physical and mental refreshment that can be received from the tea, as Sen Soshitsu (1996) points out, tea drinking was also a means of spiritual refreshment, of moving into a realm of spiritual conviviality... by drinking tea, to disencumber themselves, to rise above the vexations of the mundane, and to cross over into another dimension where they can enjoy freedom from the world's cares.

词汇表

Lu Yu 陆羽(733—804年)，字鸿渐，复州竟陵(今湖北天门)人，是唐代著名的茶学家，被尊为"茶圣"，祀为"茶神"，以著世界第一部茶叶专著《茶经》而闻名于世。他对茶叶有浓厚的兴趣，长期实施调查研究，熟悉茶树栽培、育种和加工技术，并擅长品茗。唐上元初年(公元760年)，陆羽隐居苕溪(今浙江湖州)，撰《茶经》三卷。他开启了一个茶的时代，为世界茶业发展做出了卓越贡献

Tea Classic《茶经》。中国乃至世界现存最早、最完整、最全面介绍茶的第一部专著，被誉为茶叶百科全书，唐代陆羽著

Zen Buddhism 禅宗。禅宗又名佛心宗，是中国化后的佛教。从禅宗思想体系的内涵、结构、核心来看，禅宗的基调是以心性论为基点，通过心性修持获得心性升华的心性学说，是一种摆脱烦恼、追求生命自觉和精神境界的文化理想

Buddhist 佛教徒

meditation [ˌmedɪˈteɪʃn] n. (宗教中)默想，打坐；冥想；沉思

tea tools 茶具。指采摘、制茶的工具

tea cooking 煮茶

tea ceremony 茶礼、茶艺

tea events 茶事

cake tea 饼茶。唐、宋两代是中国饼茶生产的鼎盛时代，当时又有茶饼、团茶之称。其制作方法，据陆羽《茶经·三之造》介绍，一般都经过"采之、蒸之、捣之、拍之、焙之、穿之、封之"共七道工序。宋代的"大小龙团"是一种著名的饼茶。饼茶饮用时要先敲碎，再碾细、过筛，然后用沸水冲泡盏中的茶末，要有一套专用工具

silk scroll [sɪlk skrəʊl] 绢本

sacrifice and formal ceremony 祭祀仪式

Sen Soshitsu 千宗室，日本茶道最大流派里千家第十五代家元

思考题

1. 将下面的句子译成英文。

(1) 在唐代(618—907年)，中国的辉煌时代之一，饮茶成为一种艺术。唐代陆羽(733—804年)，被尊称为"茶圣"或"茶神"。他出版了世界上第一本茶书《茶经》。

(2) 收养陆羽的僧人是爱茶之人，陆羽从小就为他准备茶。积年累月，陆羽的煎茶技巧得到了提高，对茶产生了浓厚的兴趣。后来，陆羽专心研究茶叶，并将研究成果写在他的杰作《茶经》中。

(3)《茶经》分上、中、下三卷，包括茶的本源、制茶器具、茶的采制、煮茶方法、历代茶事、茶叶产地等十章。

2. 将下面的句子译成中文。

(1) The widespread distribution of tea can be attributed to the extensive practice of Zen Buddhism in the whole country. Because sleeping and eating were strictly prohibited for Buddhists practicing meditation, they could only drink tea.

(2) This book contributed to turning tea drinking into an art and helped to popularize the art of tea drinking across the country. It played a great role in giving tea cultural significance.

(3) Besides the physical and mental refreshment that can be received from the tea, as Sen Soshitsu points out, tea drinking was also a means of spiritual refreshment, of moving into a realm of spiritual conviviality.

参考文献

[1] DAI HONGWU. A contemporary teashop design based on Chinese traditions [D]. Virginia, U.S.: Virginia Polytechnic Institute and State University, 1999.

[2] MARTIN L C. A history of tea—The life and times of the World's Favorite Beverage [M]. Vermont, U.S.: Tuttle Publishing, 2018.

Lesson 2　The Chinese Art of Tea

More than three millennia ago in China, tea evolved from a medicinal drink or vegetal food into a singular beverage. Later, over 1,200 years ago during Tang Dynasty, Lu Yu wrote *the Classic of Tea*, the first surviving tea book in existence in the world. In his book the utensils to prepare, brew, and serve tea were described. Lu Yu ranks different kinds of water based on their suitability for tea. He details how tea is best, how fire is correctly made and how water is properly boiled, he laid out much of what was required of tea as an art form. Precisely because of Lu Yu's excellent work in writing such a noble volume, the use of tea as a beverage became popularized in Tang Dynasty and subsequent eras, leading to tea becoming a common worldwide beverage.

However, not until the Song Dynasty did tea drinking become much more intertwined with art. Fast-forward to today, and Tea Art, or Cha yi, a new concept adapted from the ancients, is widely practiced and enjoyed in China by millions and is quickly increasing in popularity. In a narrow sense, Cha Yi involves only the way people appreciate tea using all the special paraphernalia and following the elaborate process and rituals associated with the artistic tea drinking rules in polite company. Therefore, the art of tea means the way of appreciating tea that requires the proper utensils, procedure and rituals. The following are the basic techniques for tea brewing.

1. Water

To make tea, the dry tea-leaves must be soaked in boiling water in order to release its flavor into the water. The quality of the water directly influences the quality of the tea. To emphasize the importance of water to tea making, Chinese say "water is the mother of tea". Chinese literature is full of poetic and imaginary descriptions of how tea and water can make a perfect match because only good water can show off the color, aroma and taste of the tea.

Modern day science confirms that before the spring water emerges from the ground, it had been filtered thoroughly and becomes crystal clear. The water then runs through a stream where it absorbs oxygen in the air to increase its oxygen capacity. It also absorbs minerals from the rocks, and thus possesses the quality and nutritious benefits of mineral water.

Ancient Chinese tea experts emphasized two important elements of water: quality and taste. Quality refers to the water's clarity as opposed to cloudiness; for good taste, the water should be live, not stagnant, and light as opposed to heavy. Water should be sweet and cool.

2. Tools

Some types of tea traditionally require the use of certain tools, shapes and materials (Figure 3.3). For example, sometimes two cups may be used. Tea is first poured into one cup, and then is poured from the first cup into the second one, so that you can enjoy the fragrance of very hot tea evaporating.

Figure 3.3　A simple modern tea sets (*From left to right*, *burner*, *flower vase*, *tasting cup*, *kettle*, *lidded tea bowl with saucer-Gaiwan*, *tea container*)

扫码看彩图

Of the many tools that appear in the preparation of Chinese tea, some are absolutely essential, while some are just for convenience and may or may not be used. For simplicity, here is a list:

The teapot—Teapots for brewing Chinese tea are usually made of clay, without any glazing, and are normally very small; a teapot the size of a fist is already a large one! You might have seen some of these in Chinese antique shops or in flea markets and thought they were toys for some sophisticated China doll: these are actually the real thing as it is much easier to control the temperature if the teapot is small.

In Chinese, the verb that means "to breed" (animals) or "to raise" (children) also applies to teapots. Tea lovers often "raise" several teapots, each for a different kind of tea. Because the clay is porous, it absorbs part of the tea, and it is said that the "oil" contained in the tea leaves will slowly build up on the inside of the teapot, to give it a healthy shine. New teapots are normally bathed in tea before the first use, to wash off the smell of the clay and to start the coating process. As a consequence, one should never wash a teapot with any kind of detergent. Just pour hot water in it, that is all it needs!

The kettle—To keep boiling water handy. It usually comes on a stand where coal (or an alcohol burner) keeps the water hot. Nowadays, electric kettles are often used.

The tea leaves container—It must be sealed tight, so tea leaves will not be exposed to humidity.

The teacups (Figure 3.4)—They come in many shapes, tall like miniature fruit-juice glasses, short and stout like little bowls, usually without an ear. Their size is proportionate to the size of the teapot, and one can usually fill four to six cups from one brew.

Figure 3.4 Tea cups (*From left to right, glass cup, small tasting cup, small tea cups with different shape and volume*)

扫码看彩图

A pot for the tea (Figure 3.5)—Because tea must not be left on the leaves more than a few minutes, and the guests might not drink quickly enough, it is handy to have a second teapot, or a small jug (Gong Dao Bei, fair cup), to pour the tea into when it is ready. Sometimes the tea is poured through a small filter, to stop leaf particles getting through.

The tea spoon (Figure 3.6)—Usually made from a small section of bamboo, split in two along its length, it looks a bit like the large spoon used in traditional western groceries to measure grain, coffee beans and flour. It is used to measure the tea leaves before putting them in the pot. It also avoids contact between the leaves and the hand; as this might spoil the taste. Sometimes a little funnel is also used to make sure all tea leaves get into the tiny teapot.

(1)　　　　　　　　　　　　　(2)

Figure 3.5　Tea brew container with different shapes and materials

(1)　　　　　　　　　　　　　(2)

Figure 3.6　Tea spoons

The used leaves container—After use, the leaf dregs should not be left in the teapot, but should be dug out and disposed of in the used leaves container, usually a deep plate.

The tongs (Figure 3.7)—This tool (like large bamboo tweezers) is used to handle the cups while "washing" them in boiling water; also used to dig leaves out of the teapot afterwards.

(1)　　　　　　　　　　　　　(2)

Figure 3.7　Tea tongs

The poker—A very thin tool made of bamboo or turtle shell or horn, used to dislodge tea leaves that could get stuck in the teapot's spout.

The teacloth (**rag**)—Lots of pouring and soaking takes place during the whole process. This small piece of cloth can be quite handy to wipe things dry.

The tea plate (Figure 3.8)—This rectangular or oval plate looks like a large flat box, on the top of which tea is prepared and served, with openings in the lid so water and spilt tea can drip through and be collected in the box. Often made of bamboo or other waterproof woods.

(1)

(2)

Figure 3.8 Tea plates

扫码看彩图

For simplicity, only the first two are essential (teapot and kettle), everything else is optional. Some tea lovers will even drink tea directly from the spout of their very tiny personal teapots!

3. Temperature

Water has to be hot, sometimes just boiling, but with very tender leaves, like green tea such as Long Jing (*dragon well*), the temperature can be as low as 60 or 70 ℃ (so as not to cook the leaves). The rule of the thumb is: the darker the leaves are, the hotter the water should be.

If the leaves are still rolled in tiny beads after the first infusion, after the water was left on them for even 2 to 3 minutes, then the water was probably not hot enough (the leaves did not "open"). If the leaves and the tea smell of cabbage, then the water was probably too hot.

4. Brewing technique

The teapot and cups are first washed in boiling water, then turned around and are left to dry for a few seconds. The teapot is then filled up to 1/4 or 1/3 with tea leaves, before the water is poured. If the tea leaves are good quality, when coming in contact with the hot steamy teapot, they will start to "exhale" their first fumes. You will notice something like the smell of freshly aroma, with a hint of a fruity or floral fragrance. After the 4th or the 5th time of adding water, the tea leaves should open to their original size and shape, and will fill the teapot completely.

Water is poured in the teapot, up to the top, and the lid is placed back on. Sometimes some water is poured on the top of the lid, to keep the teapot hot. Infusion time is flexible depending on your taste. After a minute or so, the tea is poured into the cups, and any remaining tea liquid has to be poured into an empty pot, so that no tea liquid remain on the leaves while you drink the first cup. This is very important, or it will turn bitter and spoil the leaves.

Water is poured several times on the same tea leaves before they have surrendered all their fragrance. Each time, the water is left a little longer on the leaves (adding about a minute to each turn). Between brews, do not empty your teapot completely and leave a little bit of tea liquid inside leaves to strengthen your next brew. Each brew will develop a different flavor, with stronger elements fading away to reveal more subtle ones, until the leaves are completely brewed. This may take three to seven turns, depending on the tea type and quality, and the duration of each infusion.

5. Time

When to drink a specific type of tea is not just a matter of taste or availability. In traditional Chinese medicine, non-fermented, raw, green tea is considered very "cold", and is best drunk in the middle of the afternoon, when the body's energies are at their "hottest". In general, the less roasted and fermented the tea leaves, the "colder" they are. Drinking "cold" in the morning can bring unpleasant digestive effects, and can make you actually feel cold.

It seems also that less roasted/fermented tea releases more stimulating substances (mostly caffeine and vitamin C). Legend has it that tea was first discovered by a monk from India who used to chew raw tea leaves to stay awake through his long hours of studying religious texts. If you are sensitive to caffeine, you drink this type of tea too late in the afternoon/evening and might not fall asleep easily.

词汇表

intertwine [ˌɪntəˈtwaɪn] v. 紧密相连；(使)缠结，缠绕在一起
paraphernalia [ˌpærəfəˈneɪliə] n. (尤指某活动所需的)装备，大量用品
tea art 茶艺术
Cha Yi 茶艺，茶的冲泡、品饮艺术
tea brewing 泡茶
Water is the mother of tea 水为茶之母
spring water 泉水
stagnant [ˈstæɡnənt] adj. (水或空气)不流动而污浊的；停滞的

> raise teapot 养壶。养壶是茶事过程中的雅趣之举。透过泡养摩挲的过程，茶壶以其器面的日渐温润
> tea leaves container 茶叶罐，储存茶叶器具
> teacup 茶杯。用来品饮茶汤用的杯子，有大杯，也有小杯
> A pot for the tea 茶盅，用来盛放泡好的茶汤，平均茶汤浓度，又称公道杯
> tea spoon 茶则，量取茶叶
> tea funnel 茶漏
> tea tong 茶夹
> Tea poker 茶针
> teacloth (rag) 茶巾
> tea plate 茶盘
> rule of the thumb 经验法则
> non-fermented tea 不发酵茶
> roast 烘焙
> fermented tea 发酵茶

思考题

1. 将下面的句子译成英文。

（1）正是由于陆羽撰写了如此出色的著作，茶叶作为饮料的使用在唐代及其以后的时代得到普及，使得茶成为世界性的饮料。

（2）水的质量直接影响茶汤的质量。为了强调水对茶的重要性，中国人说水是茶之母。中国文学充满诗意和想象地描述茶和水如何完美匹配，因为只有好的水可以展示茶的汤色、香气和滋味。

（3）首先将茶壶和杯子在沸水中烫洗，荡壶沥干几秒，再泡茶冲水之前将茶壶投入壶容积1/4或1/3的茶叶。如果茶叶质量很好，当茶叶与壶身热气接触，即刻散发出茶香。

2. 将下面的句子译成中文。

（1）Tea Art, or Chayi, a new concept adapted from the ancients, is widely practiced and enjoyed in China by millions and is quickly increasing in popularity.

（2）In a narrow sense, Cha Yi involves only the way people appreciate tea using all the special paraphernalia and following the elaborate process and rituals associated with the artistic tea drinking rules in polite company.

（3）Water has to be hot, sometimes just boiling, but with very tender leaves, like green tea or Long Jing, the temperature can be as low as 70 or 80 ℃ (so as not to cook the leaves).

参考文献

MARTIN L C. A history of tea—The life and times of the World's Favorite Beverage [M]. Vermont, U.S.: Tuttle Publishing, 2011.

Lesson 3　The Japanese Way of Tea(chanoyu)

1. History

"In the fifteenth century," says Okakura Kakuzo（冈仓天心）, "Japan ennobled tea into a religion of estheticism—teaism".

Previous to the Heinan period（平安时代）, 794-1159 AD, in the age of Buddhist culture in Japan, the Japanese made tea-drinking a pretext for religious and poetic conversation, but it was not until the close of the Heinan period that a regular ceremonial began to be associated with it; a ritual which contributed to the propagation of Buddhism, and the cultivation of that literary spirit which, centuries after, produced the most brilliant period of the emperor's Court and its literature.

Chanoyu means "hot water tea". The first meetings were held in the temple groves, where everything was in harmony with the solemn occasion. Transported to the towns there was an attempt to simulate natural surroundings by celebrating the ceremony in a little room set apart in a garden. Rikyu（千利休）, founder of the most popular school of Chanoyu, forbade frivolous conversation in the tea room, and demanded that the simplest movements be performed according to strict rules of ceremony and a prescribed decorum. A subtle philosophy lay behind it all, and its finished product is known to the Japanese as "Teaism". Teaism represents much of the art of Japanese life.

Teaism is a cult founded on the worship of the beautiful. Love of nature and simplicity of materials are its keynotes. It inculcates purity, harmony, mutual forbearance. It is reflected in their porcelain, lacquer, painting and literature. The nobility and the peasantry do it homage. The Japanese speak of a certain type of individual as having "no tea in him", when he is incapable of understanding the finer things of life; and the esthete is sometimes said to have "too much tea in him". Teaism has been called the art of concealing beauty that you may discover it, of suggesting what you dare not reveal; the noble secret of laughing at yourself, calmly yet thoroughly—the smile of philosophy.

Foreigners often wonder at this seeming much ado about nothing. Okakura makes answer: "When we consider how small after all the cup of human enjoyment is, how soon overflowed with tears, how easily drained to the dregs in our quenchless thirst for infinity, we shall not, blame ourselves for making so much of the teacup. Mankind has done worse. In the worship of

Bacchus, we have sacrificed too freely; and we have even transfigured the gory image of Mars. Why not consecrate us to the queen of the Camellias, and revel in the warm stream of sympathy that flows from her altar? In the liquid amber within the ivory porcelain, the initiated may touch the sweet reticence of Confucius, the piquancy of Lao – tse, and the ethereal aroma of Sakyamuni himself."(注：当我们在杯盏享尽之后，了知人类的欢愉何其渺小，眼泪的涌溢何其迅速，将无限的不熄渴望饮干喝尽，只残留时光的渣滓又何其容易，我们就不会责备自己沉湎其中了。人类不是已经做得更糟了么？对酒神巴克斯的崇拜，使我们献出太多的祭品；对战神玛尔斯的景仰，让我们抹去了他身上的斑斑血迹。那何妨伏身于茶之仙女的裙裾之下，陶醉于她甘露瓶中涓涓而出的仁爱暖流呢？从象牙白瓷杯内的琥珀色琼浆里，茶道的门徒或可一品孔子温雅含蓄，老子的辛辣快意，还有佛陀的缥缈芬芳。)

2. Basic steps

Chanoyu, also called Sado or simply Ocha in Japanese, is a choreographic ritual of preparing and serving Japanese green tea, called Matcha, together with traditional Japanese sweets to balance with the bitter taste of the tea. Preparing tea in this ceremony means pouring all one's attention into the predefined movements. The whole process is not about drinking tea, but is about aesthetics, preparing a bowl of tea from one's heart. The host of the ceremony always considers the guests with every movement and gesture. Even the placement of the tea utensils is considered from the guests view point (angle), especially the main guests called the Shokyaku (正宾). Its ultimate aim is the attainment of deep spiritual satisfaction through the drinking of tea and through silent contemplation (Figure 3.9).

(1)

(2)

扫码看彩图

Figure 3.9　Japanese tea ceremony (Chanoyu)

The ritual preparation of tea is very simple. Simplicity is one of the basics for preparing a bowl of green tea for the guests. Preparation styles can vary according to the season or the level of formality of the meeting. The simple procedures include：

Host preparation—The host needs to send formal invitations to the guests and to prepare his/her soul for the ceremony by leaving behind all worldly thoughts and just focusing on obtaining a certain harmony and equilibrium within himself or herself. The practical preparation starts with choosing the right tools depending on the season and on the part of the day the ceremony will take place. The host goes on by cleaning the tea room (garden, if it's summertime), the tools and changing the tatami (the Japanese carpets used on the floors of Japanese traditional homes). If the ceremony will also involve a meal, than the host needs to start preparing it very early in the morning.

Guest preparation—They need to purify their hearts and thoughts and leave the worldly worries behind. Before entering the tea room or garden where the ceremony will be held, the guests have to wait for the signal of the host which will announce them that the host is ready to receive them. They also need to wash their hands in an attempt to symbolically get rid of the "dust" from the outside world. After the host gives them the signal, they will enter the tea room through a small door which obliges them to bow as a sign of respect to the host and to the preparations she or he has made.

Cleaning the tools—The actual preparation of matcha doesn't start until the host brings in the tools, cleans them in front of the guests before using them. The cleaning of the tools is aesthetically done with concentration and highly graceful movements. These movements can differ from a type of ceremony to another, but what is always important is the graceful posture of the host and aesthetic value of the way things are done during the ceremony. No unnecessary movements or words are allowed during the ceremony, all the things starting with the tools and ending with the guests' behavior have to be in harmony with each other.

Preparing matcha—After the tools are perfectly clean and aesthetically displayed, the preparation of matcha begins. We have dedicated an entire article to matcha green tea, so feel free to check it out for detailed explanation regarding matcha preparation. Usually the host adds in the tea bowl three scoops of matcha per guest. After adding the powder, the hot water is also added to the bowl and the composition gets whisked into a thin paste. More water is added afterwards.

Serving matcha—The host presents the prepared tea bowl to the main guest and they exchange bowls. This first guest admires the bowl then rotates it before taking a drink. The guest wipes the rim of the tea bowl then offers it to the next guest who repeats these movements. These movements are repeated until the bowl reaches the last guest which passes it back to the host.

Completing the ceremony—After all the guests have taken a drink of tea, the host cleans the bowl. The host will also rinse and clean the tea whisk and scoop again. The guests need to inspect the tools used in the ceremony after they have been cleaned as a sign of respect and admiration for the host. They carefully and respectfully examine the utensils using a cloth when handling them with extreme caution. After this phase is over, the host gathers the tools and the guests exit with another bow to complete the ceremony.

词汇表

estheticism [esˈθetəˌsɪzm] *n.* 唯美主义；唯美体验
Teaism [tiːˈɪz(ə)m] *n.* 茶道
solemn [ˈsɒləm] *adj.* 庄严的，严肃的；庄重的；隆重的，神圣的
mutual forbearance [ˈmjuːtʃuəl fɔːˈbeərəns] 相互克制；相互容忍
no tea in him 心中无茶
too much tea in him 茶气太重
ado [əˈduː] *n.* 闲话，废话；无谓的忙乱
Confucius [kənˈfjuːʃəs] *n.* 孔子
shokyaku（日）正宾
choreographic [ˌkɒriəˈɡræfɪk] *adj.* 舞蹈设计的；编舞的
matcha 抹茶

思考题

1. 将下面的句子译成英文。

（1）僧侣们发现，茶不但使他们能节食，而且能在漫长的晚课过程中保持神志清醒。

（2）茶道是一种崇拜美的文化，其精髓是表达对自然界的崇敬和对自然界赐予我们的物质的珍爱。

（3）日本茶道是日本的一种仪式化为客人奉茶之事。

2. 将下面的句子译成中文。

（1）It was especially endeared to them by the Daruma legend, and was supposed to have great healing power. Gradually tea-drinking extended from the priests and religious orders to the laity.

（2）In a certain sense it celebrates the aristocracy of taste, yet its symbol is the "cup of humanity." It is an ever present influence in Japanese culture and for five centuries it has been a dominant force in shaping the manners and customs of the Japanese people.

（3）Teaism has been called the art of concealing beauty that you may discover it, of suggesting what you dare not reveal; the noble secret of laughing at yourself, calmly yet thoroughly—the smile of philosophy.

参考文献

[1] UKERS W. All about tea(II) [M]. New York, U.S.: The Tea and Coffee Trade Journal Company, 1935: 363-364.

[2] KAKUZO OKAKURA（冈仓天心）茶之书［M］.徐恒迦,译.北京:中国华侨出版社,2015.

[3] The Japanese Tea Ceremony［EB/OL］. http://japanese-tea-ceremony.net. 2019-05-07.

Lesson 4 Afternoon Tea

Tea became a fashionable drink for the ladies of England with the coming of Princess Catherine of Braganza, the Portuguese princess and tea devotee whom Charles Ⅱ wedded in 1662. She was England's first tea drinking queen, and it is to her favorite temperance drink as the fashionable beverage of the court in place of the ales, wines, and spirits with which the English ladies, as well as gentlemen, "habitually heated or stupefied their brains morning, noon, and night".

With a population that grew from 3,000,000 at the beginning of the seventeenth century to 5,000,000 at its close, the England of Queen Catherine's time was, throughout its greater part, "open and untamed", while enjoying the culture handed down from the Elizabethans. The English poet Edmund Waller（埃德蒙·沃勒,1606-1687 AD）wrote the first eulogy of tea "*on tea*" in English verse to honor the birthday of Queen Catherine in the year of her marriage, 1662. The poem begins:

Venus her myrtle, Phoebus has his bays;
Tea both excels, which she vouchsafes to praise.
（意译:花神宠秋色,嫦娥矜月桂;月桂与秋色,美难与茶比）

The queen's taste for may account for its selection by the directors of the East India Company as a rare and costly gift to the king in 1664. In their records for that year the following entries appear: "1664 July 1. Ordered, that the master attendant do go on board the ships now arrived (from Bantam, Java), and enquire what rarities of birds, beasts, or other curiosities there are on board, fit to present to His Majesty." On the 22nd of August the Governor, having acquainted the Court of Directors that the factors had in every instance failed the company of "such things as they wait for", he was of the opinion, if the court thought fit, that a silver case of oil of cinnamon and some good thea (tea) selected, which he hoped would be acceptable. This, we are told, the court "approved very well", and on September 30 there is an entry on the general books:

Sundry accounts Oweth to John Stannion, Secy.
Presents—For a case containing six
China bottles headed with
silver··················£ 13.0.0
More for 2 lbs. 2 oz. of thea

for His Majesty············ 4.5.0

Further impetus was given to tea drinking as a fashionable entertainment at the court of Charles II when Henry Bennet, Lord Arlington, Secretary of State, and Thomas Butler, Earl of Ossory, returned to London from The Hague in 1666, bringing in their baggage a quantity of tea which their ladies proceeded to serve after the newest and most aristocratic vogue of the Continent. At this time the Netherlands represented the pinnacle of elegance in tea serving, and every home of any consequence had its exclusive tea room.

The effect of the importation by Lord Arlington and Lord Ossory was so pronounced, thanks to the teas given by their ladies, that Jonas Hanway (1712-1786 AD), a benevolent English merchant and author of a famous attack on tea, made the statement, somewhat widely copied, that they were the first to introduce tea into England from Holland; a statement controverted by Dr. Samuel Johnson, 1709-1784 AD, in his famous reply to Hanway, when he called attention to the well-established fact that tea had been taxed since the year 1660, and had been publicly sold in London for several years before that.

At the time that Lord Arlington and Lord Ossory returned, Abbé Raynal (1713-1796 AD), a French historian, informs us: "*Tea sold in London for near seventy livres (£ 2.18.4) a pound; though it cost but three or four (from 2s. 6d. to 3s. 4d.) at Batavia.*" The price was kept up with little variation, regardless of the fact that the high cost prevented its general use. However, its vogue at court gave it added interest to the ladies; and the apothecaries of London hastened to add it to their pharmacies. In 1667, Pepys recorded in his diary: "Home and found my wife making of tea; a drink which Mr. Pelling, the potticary, tells her is good for her cold and defluxions."

The idea of afternoon tea as a meal and a social event is universally attributed to Anna Maria Stanhope, Duchess of Bedford (1783-1857 AD), wife of the seventh duke. She apparently often experienced what was commonly called "a sinking feeling" between lunch and the evening meal. Thinking that a little sustenance might help, she began drinking tea and nibbling small savory treats in the late afternoon. In the first half of the nineteenth century, luncheon was a small meal taken during the middle of the day, and dinner was often not served until eight o'clock at night. The duchess found that taking tea with a little food in late afternoon was so beneficial and pleasant that she soon began inviting friends to join her at Belvoir Castle for this small afternoon meal, around five o'clock. The menu typically included small cakes, sandwiches of bread and butter, various sweets, and, of course, tea(Figure 3.10).

This practice had proven so successful and pleasant at her summer residence that when the family returned to London in the fall, Anna continued it, inviting friends for tea and a walk in the fields (fields were still plentiful close to London, in her day). The custom caught on with others, and soon many people copied her idea. It was probably not until the middle of the nineteenth century that late afternoon tea became an established custom throughout the country, and then, still only among the well-to-do. Queen Victoria loved tea, and her enthusiasm for the afternoon tea party made it even more popular. Afternoon tea receptions were introduced at Buckingham Palace in 1865.

Figure 3.10　Afternoon tea

扫码看彩图

By the time Queen Victoria died in 1901, tea was the drink for the masses in England. Tea's importance to the lower classes was exemplified by the women in small villages (particularly in Wales) who sometimes banded together to form "tea clubs". The purpose of these clubs was to get together in the afternoon and share tea, gossip, advice, and the like. When money was scarce, they shared responsibilities as well, one woman bringing the tea, another bringing the biscuits or small breads, another bringing the teapot, and so forth.

词汇表

Princess Catherine of Braganza 布拉格的凯瑟琳公主
temperance ['tempərəns] *n.* 戒酒
ale [eɪl] *n.* 浓啤酒；麦芽酒(一种酒精含量较高的啤酒)
spirit (威士忌、杜松子酒等)烈性酒
stupefied ['stjuːpɪfaɪd] *v.* 使惊讶(或惊呆、思维不清、神志不清)
untamed [ʌn'teɪmd] *adj.* 不被驯服的；野性的
Venus 维纳斯
myrtle ['mɜːtl] 香桃木；爱神木
Phoebus ['fiːbəs] 太阳神腓比斯
vouchsafed [vaʊtʃ'seɪft] *v.* 给予，赐予
Java 爪哇

writ v. "write"的过去式和过去分词

theabe 铁皮

Secy = secretary

Impetus ['ɪmpɪtəs] n. 动力；促进；势头

earl [ɜːl] n. (英国)伯爵

Hague [heig] n. 海牙

vogue [vəʊg] n. 时尚，流行；时髦的事物；adj. 流行的，时髦的

aristocratic [ˌærɪstəˈkrætɪk] adj. 贵族的

pinnacle [ˈpɪnək(ə)l] n. 顶级

benevolent [bɪˈnev(ə)l(ə)nt] adj. 好心肠的；与人为善的；乐善好施的；慈善的

merchant [ˈmɜːtʃ(ə)nt/] n. 商人；批发商

Netherlands, Holland, Dutch 三者均指荷兰。The Netherlands 指荷兰，如同 China 指中国。Holland 也指荷兰，这是因为以前以北荷兰省和南荷兰省为主体的两个省构成了荷兰，这两个省在全世界以从事贸易闻名，久而久之，人们便以 Holland 代指荷兰，但现在这两个省只是荷兰西部国土的一部分。Dutch 荷兰人，荷兰语

Batavia [bəˈteɪvɪə] 巴达维亚；巴拉维亚

apothecary [əˈpɒθɪk(ə)rɪ] n. 药剂师

pharmacy [ˈfɑːməsɪ] n. 药房

Duchess [ˈdʌtʃəs] 公爵夫人；女公爵

duke [djuːk] n. 公爵

sustenance [ˈsʌstənəns] n. 食物，营养

well-to-do adj. 有钱的；富有的；富裕的

Buckingham Palace [ˌbʌkɪŋəm ˈpæləs] n. 白金汉宫(在伦敦的英国王室官邸)

思考题

1. 将下面的句子译成英文。

(1) 因为英国妇女与男子一样都有饮酒的习惯，而且经常喝得头脑发热昏迷不醒。在英国的王后中，凯瑟琳公主可以称得上是饮茶第一人。更让人称道的是，她让这种温和的饮料——茶取代了葡萄酒、烧酒等，成了宫廷流行的饮料。

(2) 王后对茶有很深的嗜好。因此，1644 年英国东印度公司精选茶叶作为珍贵的贡品。

(3) 当时，荷兰人的饮茶方式最为讲究，任何家庭都另辟一室专门用来饮茶。

2. 将下面的句子译成中文。

(1) On the 22nd of August the Governor, having acquainted the Court of Directors that the factors had in every instance failed the company of "such things as they wait for", he was of the opinion, if the court thought fit, that a silver case of oil of cinnamon and some good theabe selected, which he hoped would be acceptable.

(2) In 1667, Pepys recorded in his diary: "Home and found my wife making of tea; a

drink which Mr. Pelling, the potticary, tells her is good for her cold and defluxions."

(3) In the first half of the nineteenth century, luncheon was a small meal taken during the middle of the day, and dinner was often not served until eight o'clock at night.

参考文献

[1] UKERS W H. All about tea(Ⅰ)[M]. New York, U.S.: The Tea and Coffee Trade Journal Company, 1935: 43-45.

[2] MARTIN L C. A history of tea—The life and times of the World's Favorite Beverage [M]. Vermont, U.S.: Tuttle Publishing, 2018.

Lesson 5 Indian Tea Culture

Robert Fortune's first gardening job was working in the Edinburgh Royal Botanic Gardens, a position that left him with a lifelong fascination with plants. It was on his initial trip to China in 1843 that Fortune saw tea growing for the first time. In this journey, he became the first Westerner to realize that green and black tea come from the same plants. In June 1844, Fortune decided to travel to the forbidden city of Soochow (Suzhou), which was still closed to Westerners. This necessitated his traveling in disguise. He was able to pass himself off as a Chinese merchant. Fortune not only collected plants on all of his adventures, he also kept careful notes on the soils and climate, and paid particular attention to how tea was planted, harvested, and processed.

In August 1848, he made his second journey to China, sent this time by the East India Company. His goal was to find the best possible tea plants and seeds for transplanting and planting in India. He managed to collect and send twenty thousand tea plants to India, using four different ships to minimize the danger of losing all the plants to one possible catastrophe. Because he did not know whether or not the tea seeds would stay viable throughout a long journey, Fortune used the newly invented Wardian cases to germinate the seeds en route. The Wardian case, was a miniature glass greenhouse, much like a terrarium. It allowed plant explorers to immediately plant seeds they had collected, then return to their home ports with small plants, rather than bags of seeds that might not have survived many months or even years at sea. Fortune was also able to hire eight Chinese tea experts to travel with him and to purchase implements and materials needed for processing the tea. The East India Company was ecstatic with his successful forays into China, and, on the basis of what he had learned in China, Robert Fortune was able to help tea production in India increase quickly and dramatically.

By 1862, two million pounds of tea were produced in India. In 1866, the amount had grown to six million pounds. Although a great deal of tea was being produced in India, the production was very expensive, and the tea itself was of inferior quality. After a few years, the tea industry was increased production and better-quality tea, particularly from the Assam region. In 1888, eighty-six million pounds of tea were produced.

With the stages of the establishment of the British supremacy nearly all over in India, the tea culture took a visible shape, when the upper class families of the subjugated and native states started to be acquainted with the English culture and manners. In consequence the upper class families of urban areas started to imitate English culture in their respective household affairs. It is noteworthy that all the servants and staffs of the English administrators' households in India were Muslims, because the Hindus had strong caste prejudices. Muslims do not drink alcohol according to their religious beliefs and they also do not have traditional drink. In this case, it is easy for them to accept tea as a drink. Whatever the reason behind this issue, Muslim contributed for the development of tea culture in India.

It is well-known that it took long time for the Indians to be accustomed with tea generally. There is no question that Indian became accustomed to drink tea after coming in contact with British people. The habits of taking tea by the Indian renowned personnel had a direct impact of on their fellow countrymen; many of them later followed the lead without any hesitation. This in fact had helped to create a positive atmosphere for drinking tea. The custom of tea drinking was spread widely since the 20th century in India.

Hindus did not accept tea in the early stage because of caste prejudice but Hindus played vital role to form the milk tea culture in India. Hindus keep cows and worship as a god and they used to have milk from ancient time. When the fashion of tea drinking spread in India, they did not care at first. Later, they also started to drink tea but adding milk. Because tea was not tasty for them, they had accustomed to drinking milk. Tea became delicious after adding milk. Slowly and gradually, the custom of milk tea spread all over India. In this way, Hindus contributed to form milk tea culture in India.

In India the scenario of tea culture is completely different now. Tea is available nearly in every place in the cities and towns, and in some places like railway and mills, tea shops are kept open even in the mid-night. The passengers tired for the journey and also for staying up far into the night and labourers also take tea in small tea stall located nearby for removing fatigue. India is totally different culturally. In Japan, people go to Izakaya after finishing their duty and take alcohol. Even today taking alcohol is not considered as a good thing in India. So they take tea or some refreshments instead of alcohols. This is the one reason that India consumes more than 80% of its production. Tea was introduced only in the 19th century, but finally tea (popularized as *chai* in India) has turned to be the national beverage of India.

词汇表

Robert Fortune 罗伯特·福特尼，英国园艺家。他同时还是史上最大的商业间谍，曾潜入中国窃取茶产业技术

botanical [bəˈtænɪkl] *adj.* 植物学的

catastrophe [kəˈtæstrəfi] *n.* 灾难；不幸事件

Wardian case 沃德箱（培育蕨类植物等的玻璃容器）

terrarium [teˈreəriəm] *n.* 生物育养箱；玻璃花园

Burmese [bɜːˈmiːz] *n.* 缅甸人；缅甸语 *adj.* 缅甸的；缅甸人的；缅甸语的；缅甸文化的

forays [fɒreɪ] *n.* 突袭

supremacy [suːˈpreməsi] *n.* 最大权力；最高权威；最高地位

subjugate [ˈsʌbdʒugeɪt] *v.* 使屈服；使服从

Izakaya（日本）居酒屋

思考题

1. 将下面的句子译成英文。

（1）1848年，福琼接受东印度公司的派遣，深入中国内陆茶乡，将中国茶树品种与制茶工艺引进东印度公司开设在喜马拉雅山麓的茶园，结束了中国茶对世界茶叶市场的垄断。

（2）1862年印度开始大量生产茶叶，但茶叶品质低劣。

（3）随着英国统治的加强，英属印度的上层开始接触并熟悉英国饮茶文化，推动了饮茶文化在印度的普及，以至今天印度以茶叶为国饮。

2. 将下面的句子译成中文。

（1）Muslims do not drink alcohol according to their religious beliefs and they also do not have traditional drink. In this case, it is easy for them to accept tea as a drink.

（2）Hindus keep cows and worship as a god and they used to have milk from ancient time. When the fashion of tea drinking spread in India, they did not care at first. Later, they also started to drink tea but adding milk.

（3）Even today taking alcohol is not considered as a good thing in India. So they take tea or some refreshments instead of alcohols.

参考文献

［1］UKERS W H. All about tea(I)［M］. New York, U.S.: The Tea and Coffee Trade Journal Company, 1935:135-137.

［2］MARTIN L C. A history of tea—The life and times of the World's Favorite Beverage

[M]. Vermont, U.S.: Tuttle Publishing, 2018.

[3] GURUNG R. Formation and expansion of tea culture in India—with a special reference to Bengal [J]. Journal of East Asian Cultural Interaction Studies, 2012, 5: 449-471.

Lesson 6   Russian Tea Culture

Tea is part of Russian culture and about 82% of Russian consumes tea daily. Since 1638, tea has had a rich and varied history in Russia. Due in part to Russia's cold northern climate, it is today considered the *de facto* national beverage, one of the most popular beverages in the country, and is closely associated with traditional Russian culture. It was traditionally drunk at afternoon tea, but has since spread as an all day drink, especially at the end of meals served with dessert.

An important aspect of the Russian tea culture is the ubiquitous Russian tea brewing device known as a samovar, which has become a symbol of hospitality and comfort. Samovars were first made in Russia in the early 1770s, they became somewhat commonplace in the 1890s. But luckily for tea fanciers, the samovar did become the four-legged, fanciful, ornate urn that will forever be the icon of Russian tea culture.

Samovars are elaborately designed vessels that boil water and keep it heated to the proper tea-brewing temperature. Water is filled from the top and heated and stored in the middle section. Today most samovars are electric, but in the past a center tube running up the middle of the samovar held charcoal or hot coals necessary for heating the water (Figure 3.11). At the top of the samovar, a crownlike top (*komforka*) is made to hold a small metal or porcelain teapot (*chainik*) and keep the contents warm. In folk tradition, the *chainik* is sometimes covered with a tea cozy doll made in the folk-art likeness of an old woman; her abundant skirt billows out and covers the teapot, keeping the heat in. The *chainik* contains strong concentrated tea known as *zavarka*. When it is time for a cup of tea, the user pours a small amount of *zavarka* into a teacup or a straight-sided tea glass that sits in a fanciful filigree silver or enamel holder with an attached handle (*podstakannik*). Hot boiled water (*kipyatok*) is released from the spigot, and the tea is diluted to taste. The usual ratio is ten parts *kipyatok* to one part *zavarka*, but adjustments are always made for individual taste.

Muscovites have long been tea drinking connoisseurs. They were the first Russians to taste the beverage. This was in 1638, when an ambassador brought Czar Alexei Mikhailovich 130 pounds of tea from Mongolia. The Czar sent the Mongolian khan a hundred sableskins as a token of his gratitude. In 1679, Russia concluded a treaty on regular tea supplies from China via camel caravan in exchange for furs. The Chinese ambassador to Moscow made a gift of several chests of tea to Alexis I. However, the difficult trade route made the cost of tea

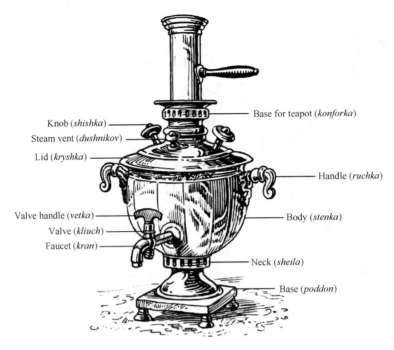

Figure 3.11　A classic Samovar

extremely high, so that the beverage became available only to royalty and the very wealthy of Russia. In 1689, the Treaty of Nerchinsk (尼布楚条约，俄方称涅尔琴斯克条约) was signed that formalized Russia's sovereignty over Siberia, and also marked the creation of the Tea Road that traders used between Russia and China.

Between the Treaty of Nerchinsk and the Treaty of Kiakhta (1727 年，恰克图条约), Russia would increase its caravans going to China for tea, but only through state dealers. In 1706, Peter the Great made it illegal for any merchants to trade in Beijing. In 1736, Catherine the Great established regular imports of tea. By the time of Catherine's death in 1796, Russia was importing more than 3 million pounds by camel caravan in the form of loose tea and tea bricks, enough tea to considerably lower the price so that middle and lower class Russians could afford the beverage.

Tea did not become available or affordable to the vast majority of Russia's population until the late nineteenth century, but so quickly and thoroughly was the drink incorporated into that country's social and cultural life that even today most Russians believe their tea traditions are far older. The peak year for the Kiakhta tea trade was in 1824, and the peak year for the tea caravans was 1860. From then, they started to decline when the first leg of the Trans-Siberian Railway was completed in 1880. Faster train service allowed for tea to be imported from nearly a year and a half to eventually just over a week. The decline in Chinese tea in the mid 19th century in turn meant that Russia began to import more tea from Odessa, and London. By 1905, horse drawn tea transport had ended, and by 1925 caravan as the sole means of transport for tea had ended.

Neither tea nor the samovar is Russian in origin. However, Russian tea traditions, epitomized by the samovar (Figure 3.11), were burned into the popular memory in the nineteenth century. Beginning with Aleksander Pushkin (1799-1837 AD), the great Russian literary figures of the nineteenth century helped virtually institutionalize the rituals, traditions and taboos surrounding tea drinking, thereby making them distinctively Russian. Pushkin's poetry is full of rich descriptions of the everyday habit of tea drinking among the Russian upper classes, and his writings were among the first to establish the ritual of *chaepitie* (tea-drinking) in the Russian national consciousness.

Tea in Russia was not regarded as a self-dependent beverage; thus, even the affluent classes adorned it with a jam, syrup, cakes, cookies, candies, lemon and other sweets. Within Russia, tea preparation differs, but usually includes lemon, and sugar or jam. Tea sachets are widely popular, but when a teapot is used it is very common to make a strong brew, then pour some into a cup and top it with hot or boiling water, adding milk and sugar afterwards. Traditional forms of Russian tea ware include the Russian tea brewing urn called a samovar, the Lomonosov tea sets (Figure 3.12), and traditional Russian tea glass holders.

Figure 3.12 Contemporary blue-and-white Lomonosov porcelain tea wares are joined by decorative Russian siliver tea glass holders (podstakanniks) outfitted with handle-less tea glasses

扫码看彩图

词汇表

de facto [ˌdeɪ ˈfæktəʊ] *adj.* 实际上存在的(不一定合法) *adv.* 实际上, 事实上
samovar [ˈsæməvɑː(r)] *n.* 俄式茶炊
tea fancier 茶叶爱好者

> billow [ˈbɪləʊ] v. 鼓起；(烟雾)涌出
> filigree [ˈfɪlɪɡriː] n. 金银丝饰品
> enamel [ɪˈnæml] n. 搪瓷；釉质；瓷釉
> spigot [ˈspɪɡət] n. (龙头中的)塞，栓
> Muscovite 莫斯科人
> connoisseur [ˌkɒnəˈsɜː(r)] n. 鉴赏家；行家
> Czar [zɑː] 沙皇
> sableskin 貂皮
> camel caravan [ˈkæml ˈkærəvæn] 骆驼商队
> epitomize [ɪˈpɪtəmaɪz] v. 成为…的典范(或典型)
> Aleksander Pushkin 亚历山大·普希金(1799—1837年)，是俄国著名文学家、诗人、小说家，现代俄国文学的创始人，19世纪俄国浪漫主义文学主要代表，同时也是现实主义文学的奠基人，被誉为"俄国文学之父""俄国诗歌的太阳""青铜骑士"，代表作有《自由颂》《致恰达耶夫》《致大海》等
> affluent classe 富裕阶层
> jam [dʒæm] n. 果酱
> sachet [ˈsæʃeɪ] n. (塑料或纸质)密封小袋；(置于衣物中的)小香囊，小香袋
> urn [ɜːn] n. 大茶壶；瓮，缸

思考题

1. 将下面的句子译成英文。

(1) 在19世纪的俄国人用茶炊喝茶是一种普遍的事情，茶炊慢慢变成了俄罗斯人家里必备的东西，后来变成了一种俄罗斯传统喝茶仪式，也开始正式属于一种俄罗斯独特的文化。

(2) 从亚历山大·普希金(1799—1837年)开始，19世纪的俄国文学家促使俄罗斯饮茶文化发展出自身特色。

(3) 随着1880年跨西伯利亚铁路的建成，传统中俄万里茶道(Sino-Russo Tea Road)上的茶叶运输骆驼商队逐渐消失。

2. 将下面的句子译成中文。

(1) An important aspect of the Russian tea culture is the ubiquitous Russian tea brewing device known as a samovar, which has become a symbol of hospitality and comfort.

(2) Samovars are elaborately designed vessels that boil water and keep it heated to the proper tea-brewing temperature.

(3) Hot boiled water (*kipyatok*) is released from the spigot, and the tea is diluted to taste. The usual ratio is ten parts *kipyatok* to one part *zavarka* (strong concentrated tea), but adjustments are always made for individual taste.

参考文献

［1］YODER B. Making tea Russian：The samovar and Russian national identity，1832-1901［D］. Ohio，U. S. ：Mianmi University，2009.

［2］World Heritage Encyclopedia. Russian tea culture［M/OL］. http：//www.worldlibrary.org/articles/eng/Russian_tea_culture.

［3］HEISS M L，Heiss R J. The story of tea a cultural history and drinking guide［M］. New York，U.S.：Ten Speed Press，2007：476-480.

［4］TRAPEZNIK A. The manufacturing process of samovar production in Tula［J］. International Journal for the History of Engineering &Technology，2014，84(2)：274-293.

UNIT Four Tea Cultivation and Breeding

Lesson 1 Botanical Classification of Tea

The cultivated plant species *Camellia sinensis* [(L.) O. Kuntze] is the source of the raw material from which the popular tea beverage is processed. The species is now cultivated commercially in Asia, Africa and South America. Major producers of the crop include China, India, Kenya, Sri Lanka and Indonesia. Although the crop is cultivated in many countries, there are several different types of tea plant, each with its own identifiable character and potential for unique cup quality. Because of this diversity, it is important that the different types of tea plant can be distinguished and be classified.

1. Classification in *Camellia*

The tea plant (*Camellia sinensis*) from which the beverage tea is processed, is placed in the genus *Camellia*. The genus has over 200 species and is largely indigenous to the highlands of southwest China and north eastern India. Tea is, however, the most important of all *Camellia* spp. both commercially and taxonomically. Tea was initially classified as *Thea sinensis* by Linnaeus (Linnaeus, 1753). Following the discovery, two distinct taxa were identified and classified by Masters (1844) as *Thea sinensis* (the small-leaved China plant) and *Thea assamica* (the large-leaved Assam plant). Today tea is botanically referred to as *Camellia sinensis* (L.) O. Kuntze, irrespective of species-specific differences. *Camellia sinensis* is classified under section *Thea* along with 18 other species.

At the species level, several intergrades resulting from unrestricted intercrossing between disparate parents have been documented, but have not been assigned the status of separate species. However, three distinct tea varieties have been identified on the basis of leaf features like size, pose and growth habit. These are the China variety, *Camellia sinensis*, var. *sinensis* (L.); the Assam variety, *Camellia sinensis* var. *assamica*; and the southern form also known as the Cambod race, *C. assamica* ssp. *Lasiocalyx*. The three main taxa can be differentiated by foliar, floral and growth features (Tables 4.1 and 4.2) and by biochemical affinities. Research has shown that cultivated tea is an out-crosser with an active prezygotic gametophytic late-acting self-incompatibility (LSI) system. Because of its out-breeding nature and, therefore, high heterogeneity, most cultivated teas exhibit a cline extending from extreme China-like plants to those of Assam origin. Intergrades and putative hybrids between *C. assamica* and *C.*

Table 4.1 Criteria used for differentiating two major tea varieties and sub-varieties of *Camellia sinensis*

Variety	Sub-Varieties	Growth Habit	Leaf Characteristics	Leaf Pose	Leaf Angle
China *Camellia sinensis* var. *sinensis* (L.)	*C. sinensis* var. *sinensis* f. *parviflora* (Miq) Sealy *C. sinensis* var. *sinensis* f. *macrophylla* Sieb (Kitamura)	Dwarf, shrub-like, slow growing	Small, erect, narrow, serrate, dark green in colour	Erectophile (directed upwards)	<50°
Assam *Camellia sinensis* var. *assamica* (Masters) Kitamura		Tall, tree stature, quick growing	Large, horizontal pose, broad, mostly non-serrated, light green in colour	Planophile (horizontal)	<70°

Table 4.2 Types of tea differentiated on the basis of foliar characteristics

Morphological Characteristics	Leaf Type				
	1	2	3	4	5
Mean area of an individual leaf (cm^2)	11.8	37.2	65.0	152.4	46.0
Length/breadth ratio	2.5	2.3	2.0	2.4	2.8
Mean internode length below plucking surface (cm)	1.7	3.4	5.0	6.6	3.0
Mean leaf angle from vertical (degrees)	26	29	87	99	28
Angle formed by halves of lamina (degrees)	105	120	162	189	161
Patina	Matt	Matt	Glossy	Highly glossy	Glossy
Leaf area index of mature bush	8.55	5.34	4.22	3.64	4.38

1 = Extreme China

2 = Typical between Assam and China

3 = Typical between Assam and China

4 = Extreme Assam

5 = Close to 2

[Scheme as used by Hadfield (1974)]

sinensis can themselves be arranged in a cline of specificity. Indeed because of the extreme hybridizations between the three tea taxa, it is debatable whether archetype (original) *C. sinensis*, *C. assamica* or *C. assamica* ssp. *lasiocalyx* still exists. However, the numerous tea hybrids currently available are still referred to as Assam, Cambod or China depending on their morphological proximity to the main taxa.

2. Morphological traits

Though the tea plant can grow to an under story tree or even a tree, cultivated plants are maintained as a low bush in a continuous vegetative phase of growth by cyclic pruning every three to five years to form a plucking table. Because of this cultural practice, it has become essential that vegetative characteristics be used to describe and differentiate tea taxa. However, because of its outbreeding nature and high heterogeneity, most of its vegetative, biochemical and physiological characteristics show continuous variation and high phenotypic plasticity. Despite this, leaf macromorphological features, e.g. leaf color, size and pose; leaf angle and leaf area index; have been widely used as descriptors in tea taxonomy, although chemotaxonomy and molecular (DNA) taxonomy, have also been applied. Traditionally, leaf and floral morphology, and growth habit have been the more important criteria for assigning taxonomic categories within *Camellia*. The standardized descriptor list of the International Plant Genetic Resources Institute (IPGRI) in 1997 also includes such novel methods as chemical profiling, e.g. for catechin content and Terpene Index; biochemical markers, e.g. isoenzymes; cytological markers, e.g. chromosome number, meiosis chromosome associations, identified and sequenced genes; and the more esoteric molecular marker fingerprint profiles, e.g. restriction fragment length polymorphisms (RFLP), random amplified polymorphic DNA (RAPD), amplified fragment length polymorphism (AFLP), simple sequence repeats (SSRs), sequence tag sites (STS), and expressed sequence tags (EST), etc.

The vegetative characteristics generally used in assigning taxonomic categories in tea include: leaf size (leaf length, leaf breadth), ratio of leaf length to breadth, internode length, length and girth of bud, petiole length, ratio of apical length, angle between leaf tip and axis, leaf margin, shoot density, etc. Although these related characteristics may overlap and show continuous variation, restricting their usefulness in characterization somewhat, they continue to be widely used by tea growers and scientists all over the world. Based on growth habit and leaf features, the major varieties of tea have been distinguished as China and Assam varieties, as shown in Table 4.1.

On the basis of leaf angle, for example, can tea be classed as erectophile (leaf angle < 50°), planophile (leaf angle >70°), or oligophile (leaf angle 50°–70°). While China and Assam types generally cover the erectophile and planophile plants respectively, the oligophiles are far too distinct to come under any of the two main varieties. Despite this limitation, leaf features continue to be widely used to classify tea and to study the extent of variation between cultivated taxa.

Leaf anatomical differences and particularly the variation in distribution and morphology of sclereids in leaf lamina have also been useful for differentiating tea varieties. Cambod clones have numerous sclereids compared to Assam and China clones. Similarly, sclereids in Cambod tea have wider lumen when compared to those of the other varieties. Leaf trichome hairs have also been used to identify some tea taxa. Stomata, leaf anatomical features and stomatal conductance have also been used to characterize tea clones.

3. Cytological markers

The main varieties of tea are diploids ($2n=30$), while a few cultivated tea forms a stable polyploidy series. Natural triploids, tetraploids and aneuploids have been sampled in tea populations in Japan, India and Kenya, but are reportedly present in very low number. Generally, rooting ability, leaf size and dry-weight of polyploids as well as total polyphenol content are higher than those of diploids. Pollen viability and fertility of triploids are, however, usually poor; with tetraploids being more fertile than the triploids, but less than the diploids. Studies carried out on different tea cultivars have revealed that chromosome complements of tea comprise of near metacentric to submetacentric chromosomes. Nucleolar number in tea has been demonstrated to correspond to multiples of the somatic cell number and is a good marker for ploidy in the species. Cytological investigations in tea have, however, been restricted mostly to determination of chromosome number rather than explaining the cytological basis of species differentiation.

4. Chemical/biochemical traits

Chemical profiling has also been used to characterize species within the genus *Camellia*. Tea has been demonstrated to contain characteristic compounds, such as caffeine, catechins and theanine. Only species of section *Thea* in the entire genus of *Camellia* contain galloyled catechins such as (-)-epicatechin gallate (ECG) and (-)-epigallocatechin gallate (EGCG) and the non-protein amino acid, theanine. Anthocyanin-rich tea has also been described in Japan and Kenya. Tea cultivars can also be classified based on their capacity to undergo auto-oxidation as a consequence of different polyphenol oxidase (PPO) activity.

5. Molecular markers

Molecular markers have provided an important set of descriptors which are able to differentiate between genotypes, variety types and even different species of the genus *Camellia*. These markers are highly discriminative and can even distinguish genotypes that cannot be distinguished using morphological traits. They are useful for the determination of differentiation and relatedness within cultivated tea. Molecular markers have also been demonstrated to be taxonomically informative at the species and genus level. Phylogenetic relationships in *Camellia* which have been constructed using molecular data have been similar to those obtained using morphological and biochemical affinities. Moreover, molecular tools have been utilized to generate genetic linkage maps of tea and to identify genes which are responsible for important agronomic and biochemical traits.

词汇表

taxonomy [tæk'sɒnəmi] *n.* 分类学

genetic diversity 遗传多样性

morphological [ˌmɔːfə'lɒdʒɪkəl] *adj.* 形态学的

anatomical [ˌænə'tɒmɪkl] *adj.* 解剖学的

disparate ['dɪspərət] *adj.* 由不同的人(或事物)组成的；迥然不同的；无法比较的

prezygotic [prezaɪ'ɡɒtɪk] 合子形成前的

gametophytic [ɡəˌmiːtəʊ'fɪtɪk] 配子体的

heterogeneity [ˌhetərə'dʒəˈniːəti] *n.* 异质性，不均匀性，不纯一性

species-specific ['spiːʃiːz spə'sɪfɪk] 物种特性

self-incompatibility system 自交不亲和性。指具有完全花并可以形成正常雌、雄配子，但缺乏自花授粉结实能力的一种自交不育性

patina 表面光泽

matt *adj.* 无光泽的；不光滑的

glossy *adj.* 有光泽的；光滑的

morphological trait 形态特征

vegetative ['vedʒɪtətɪv] *adj.* 有关植物生长的；植物的

vegetative reproduction 营养繁殖，无性繁殖

plucking table 采摘面

leaf area index 叶面积指数

chemotaxonomy *n.* (动植物) 化学分类学

terpene index 萜烯指数。是指茶叶中单萜烯醇及其氧化物芳樟醇与香叶醇的比值，1981 年由日本学者竹尾忠一提出。萜烯指数可作为茶树品种的化学分类指标，应用于茶树种质资源的传播途径和茶叶香气特征方面的研究。萜烯指数值高，则香气馥郁怡人；萜烯指数低，则香气高锐

isoenzymes 同工酶

chromosome number ['krəʊməsəʊm 'nʌmbə(r)] 染色体数

meiosis [maɪ'əʊsɪs] chromosome associations 减数分裂染色体关联

esoteric [ˌesə'terɪk] *adj.* 难解的，深奥的；机密的

restriction fragment length polymorphisms (RFLP) 限制性片段长度多态性

random amplified polymorphic DNA (RAPD) 随机扩增多态 DNA

amplified fragment length polymorphism (AFLP) 扩增片段长度多态性

simple sequence repeats (SSRs) 简单重复序列

sequence tag sites (STS) 序列标签位点

expressed sequence tags (EST) 表达序列标签
leaf breadth 叶宽
ratio of leaf length to breadth 叶片长宽比
internode length 间节长度
petiole length 叶柄长度
shoot density 枝条密度
sclereids 厚壁细胞
trichomes [ˈtrɪkəʊm] 表皮毛；表皮毛细胞
leaf trichome hairs 叶片茸毛
stomata [ˈstoʊmətə] n. (stoma 的复数) 气孔
pollen viability 花粉活力
species and genus 种和属
phylogenetic [faɪləʊdʒəˈnetɪk] adj. 系统发育的；系统发生的
biochemical affinities 生化亲缘关系
stress tolerance traits 抗逆性
cold hardiness 抗寒性
drought resistance 抗旱性
disease and pest resistance 抗病虫害

思考题

1. 将下面的句子译成英文。

（1）在"种"这一层面，茶叶分类学曾一直未能引起人们的关注和兴趣，直到茶叶的重要经济价值被发现。

（2）三个主要的茶种在形态学特征、生理学特征和叶片生化性状以及分子生物学特性方面存在明显差异。

（3）茶树是多年生常绿木本植物，我国通常将茶树品种按树型、叶片大小和发芽迟早三个主要性状分为三个分类等级，作为茶树品种分类系统。

2. 将下面的句子译成中文。

（1）Though the tea plant can grow to an under tree or even a tree, cultivated plants are maintained as a low bush in a continuous vegetative phase of growth by cyclic pruning every three to five years to form a plucking table.

（2）However, because of its outbreeding nature and high heterogeneity, most of its vegetative, biochemical and physiological characteristics show continuous variation and high phenotypic plasticity.

（3）Although the crop is cultivated in many countries, there are several different types of tea plant, each with its own identifiable character and potential for unique cup quality.

参考文献

WACHIRA F N, KAMUNYA S, KARORI S, *et al*. The tea plants: botanical aspects [M]//Preedy V R Tea in health and disease prevention. San Diego, U.S.: Academic Press, 2013: 1-17.

Lesson 2　Tea Cultivation

1. Cultivation

Tea has been cultivated for at least 1,500 years. Tea can be grown in many regions, but to cultivate it commercially, the three main requirements of soil, temperature, and most crucially, rainfall preclude a large number of regions and countries. Cultivation of tea is possible in areas that receive over 120-150 cm of rain annually, and have temperatures of 12-30 ℃. An ambient temperature in the range of 18-25 ℃ is thought to create optimal conditions. Optimum growing conditions are 250~300 cm annual rainfall and average temperatures of 18-20 ℃. Cultivation may occur from sea level to 2,200 meters, with some tea cultivars found as high as 3,000 m. Higher altitudes are often associated with higher tea quality. At least five hours of direct or 11 hours of indirect sunlight daily are required for tea cultivation. Soils must be well drained, sandy, thoroughly aired, deep and nutritious with a healthy layer of humus and low pH. Tea is most commonly found growing in a 4.6-6.2 band of pH values. There are notable exceptions with good teas being produced in northern and southern India and in East Africa in soils with pH values approaching 7. Drought, water logging, excessive heat and frost are harmful for the growth of tea plants and may result in a lower quality product in terms of chemistry, taste, aroma and bioactivity. Tea plants are often raised in controlled nursery conditions or other protected conditions for their first two to four years. They are classified as immature at this time and are not harvested. Once tea plants mature, they are transplanted to the field and are ready for harvest.

Generally speaking, after 50 years of bearing leaves, bushes tend to decline in terms of yield and are often replaced, but the life span of the cultivated crop is long and many thick stemmed plants are capable of yielding leaves of high enough quality to warrant maintaining them for 100 or even 120 years. There are 1,700-year-old tea trees today in Yunnan that continue to flourish and bloom.

2. Shrub formation

The majority of tea plants cultivated globally are grown in compact rows and are pruned as roughly rectangular shrubs at a height of 1-1.2 m in what is termed a picking or plucking table.

Formation of a plucking table serves to expedite harvest and to maximize yields per area. Young tea plants are usually laterally cut approximately 23 cm above the soil to encourage lateral growth and formation of a shrub. Pruning is done throughout the year, between harvest seasons or annually to encourage growth and maintain the form of the plucking table. Deep pruning occurs every four to five years.

3. Fertilizer

The tea plant loses nutrients through erosion, leaching and sustained harvesting of the crop. When soil has inadequate levels of nutrients to maintain tea bush growth it can be replenished through applications of fertilizers, organic manure and mulching of the roots. Nitrogen (N) is one of the most valuable nutrients required by the soil for producing tea as it maintains the health of the bush and significantly contributes to volume of leaf and yield rate. Fertilizers, therefore, are mainly nitrogen based, applied as sulphate of ammonia and urea.

The greatest proportional loss of elements from the soil occurs with nitrogen and potassium (K), while the loss of phosphorus (P) is comparatively small. The acidic nature of tea soils results in low levels of potassium. The loss of potassium in the tea crop is as high as 2-4 kg per 100 kg crop. A lack of potassium will affect the ability of the bush to recover from pruning. The secondary nutrients include magnesium (Mg), calcium (Ca) and sulphur (S). The absorption of magnesium by the tea plant is relatively low but its application to the soil is beneficial as it has a detoxifying effect and helps the uptake of nitrogen and phosphorus. Calcium contributes to the all important production of active growth points.

Other nutrients required in trace amounts include zinc (Zn), manganese (Mn), boron (B), iron (Fe), copper (Cu), molybdenum (Mo) and chlorine (Cl). Of these trace elements, zinc and copper deficiencies pose the greatest problems. Good fermentation results are dependent on adequate levels of copper. Zinc deficiency will affect the development of young shoots and is best rectified by application in foliar form. By the same token an excess of some trace elements can be equally deleterious inducing plasmolysis, which may prove fatal to the plant.

Excessive amounts of fertilizer will lead to a lower-quality end product and degrade soil rendering it unsuitable for future use. There are three basic strategies available to the grower. Firstly, experimentation with fertilizer levels until the desired result is achieved; secondly, using fertilizer as a direct response to crop deficiency; thirdly, evaluation of soil fertility to establish the exact nature of fertilisation required. Chemical analysis of either soil or plant can help achieve this.

4. Shading

Because the indigenous tea was found in the understorey of the forest it was assumed originally that the ideal environment for tea was under shade. Most tea was therefore planted with shade. Use of covering cultivation or planting shading trees to shade tea (*Camellia sinensis* L.) trees to produce high-quality, high-priced green tea has recently increased in China and Japan. Knowledge of shading effects on morphological and color traits and on chemical components of new tea shoots is important for product quality and productivity. However, it was found that most flavonoid metabolites (mainly flavan-3-ols, flavonols and their glycosides) decreased significantly in the shading treatments, while the contents of chlorophyll, β-carotene, neoxanthin and free amino acids, caffeine, benzoic acid derivatives and phenylpropanoids increased.

5. Pests and disease

Tea plants are subject to infestation by various pest and diseases throughout the year, causing an average 6%-55% loss in tea production. Tea plants have a number of naturally occurring predators including the short-hole borer beetle (*Xyleborus fornicatus*), red spider mite (*Oligonychus coffeae*), coccinellid and staphylinid larvae, lace-wing larvae, and Typhlodromus and Phytoseius mites. Some tea pests such as the short-hole borer beetle cause severe damage to the frame of tea plants. For some green tea types, infestation by particular pests is favorable for the sensory properties of the tea products. Insecticides have conventionally been used to control tea pests. Tea plants are also vulnerable to blister blight disease from the fungus Exobasidium vexans. This fungus is eliminated by direct exposure to sunlight and thus shade trees are removed in areas that are susceptible to infestation. Copper-based fungicides are also used to eliminate to blister blight disease.

6. Weeds

It has been estimated that on a global scale weeds account for an annual tea crop loss of 14%-15%. This is partly brought about through competition for nutrients, light and water by the weeds and partly due to the fact that weeds can harbour diseases and infestations of pests. Weed management uses several strategies to combat weed growth, although it must be stressed that not all casual weeds that grow on land under cultivation are harmful and therefore they should not be removed automatically. Providing weeds are non-competitive, their retention can greatly benefit the soil in the wet season by limiting soil wash. The removal of weeds is done about every two months and should be carried out before they bear seeds. Light weeding is sufficient to remove annual dicotyledons with their high water intake. Their perennial counterparts require more vigorous extraction.

In order to suppress the germination of weed seeds, cover crops and mulching are further effective means of counteracting weed growth. Seedling tea, which lacks the continuous cover provided by clonal tea, is more prone to weed growth, as the wider spacing of bushes cannot prevent the sun penetrating to the soil level. Likewise, weeds flourish in tea fields when tea plants are young and after bushes have been pruned. The formation of a dense plucking table will greatly minimise the necessity to weed and promote a more liberal growth of surface feeder roots. Moreover, soil disturbance will be kept to a minimum, allowing prunings and leaf-fall to accumulate to form valuable mulch.

An integrated weed management policy combining manual and chemical (herbicide) control is called for to obviate any horticultural imbalance. One natural method of keeping weeds under control that is worthy of exploration is the ability of some species of plants to exude material from their roots which inhibits the development of others.

词汇表

tea cultivation 茶树栽培
high altitude 高海拔
humus ['hju:məs] n. 腐殖质
water logging ['wɔːtə(r) 'lɒgɪŋ] 水涝, 渍水; 水淹
picking or plucking table 采摘面
lateral growth 横向生长
pruning n. 修剪
fertilizer ['fɜːtəlaɪzə(r)] n. 肥料, 化肥
leaching ['liːtʃɪŋ] v. 过滤; 滤去
organic manure [ɔːˈgænɪk məˈnjʊə(r)] n. 有机肥料
mulch [mʌltʃ] n. 覆盖物; v. 用覆盖物覆盖(土壤或根部)
foliar ['fəʊliə(r)] adj. 叶的; 叶状的 n. 叶面肥
plasmolysis [plæzˈmɒlɪsɪs] n. 质壁分离; 胞浆分离; 胞质分离
soil fertility 土壤肥力
shading 遮阳
shading tree 遮阳树
pest and disease 病虫害
predator ['predətə(r)] n. 捕食性动物
short-hole borer beetle 短孔钻甲虫
red spider mite 红蜘蛛螨
coccinellid [kɒkˈsaɪnelɪd] n. 瓢虫
staphylinid [stæfɪˈlɪnɪd] n. 隐翅虫
larvae ['lɑːviː] n. (larva 的复数)(昆虫的)幼虫
lace-wing 花边翼
Typhlodromus 飞虱
Phytoseius 植绥螨属
pesticide ['pestɪsaɪd] n. 农药, 杀虫剂
insecticide [ɪnˈsektɪsaɪdz] n. 杀虫剂
blister blight ['blɪstə(r) blaɪt] 茶饼病
Exobasidium vexans 茶饼病菌

fungicide [ˈfʌŋgɪsaɪd] *n.* 杀菌剂
dicotyledon [ˌdaɪkɒtɪˈliːdən] *n.* 双子叶植物
herbicide [ˈhɜːbɪsaɪd] *n.* 除草剂

思考题

1. 将下面的句子译成英文。

（1）在种植茶树时，种植土壤的可耕作层要含有丰富的有机质，土质疏松，有着较强的通透性与保水保肥能力，土壤的 pH 以 5~6 为宜。

（2）遮阳处理对茶树新梢的组织结构、颜色以及化学成分有重要影响。一般而言，适度遮阳可显著降低夏暑季茶树冠层温度，升高相对湿度，促进茶树生长。遮阳可降低新梢中茶多酚含量，提高氨基酸含量，有利于茶品质的形成。

（3）我国茶园发生的病虫害主要有茶小绿叶蝉、黑刺粉虱、灰茶尺蠖、茶尺蠖、茶毛虫、茶橙瘿螨以及茶饼病和炭疽病等。

2. 将下面的句子译成中文。

（1）Knowledge of shading effects on morphological and color traits and on chemical components of new tea shoots is important for product quality and productivity.

（2）Drought, water logging, excessive heat and frost are harmful for the growth of tea plants and may result in a lower quality product in terms of chemistry, taste, aroma and bioactivity.

（3）An integrated weed management policy combining manual and chemical (herbicide) control is called for to obviate any horticultural imbalance. One natural method of keeping weeds under control that is worthy of exploration is the ability of some species of plants to exude material from their roots which inhibits the development of others.

参考文献

[1] AHMED S, STEPP J R. Green tea: the plants, processing, manufacturing and production [M]// Preedy V R. Tea in health and disease prevention. San Diego, U.S.: Academic Press, 2013: 19−23.

[2] BHATTACHARYYA A, KANRAR B. Diversity of pesticides in tea [M]//Preedy V R. Tea in health and disease prevention, San Diego, U.S.: Academic Press, 2013: 1491−1501.

[3] HALL N. The tea industry [M]. Philadelphia, U.S.: Woodhead Publishing, 2000: 42−97.

[4] ZHANG Q, SHI Y, MA L, et al. Metabolomic analysis using ultra−performance liquid chromatography−quadrupole−time of flight mass spectrometry (UPLC−Q−TOF MS)

uncovers the effects of light intensity and temperature under shading treatments on the metabolites in tea [J]. Plos One, 2014, 9(11):e112572.

Lesson 3 Tea Breeding

Tea is an important revenue source for the tea producing countries in the world, including China. The strong interdependence of countries with respect to tea production and consumption has led to its high value in the international marketplace – and to the need for significant investment in breeding new cultivars. The development of new tea cultivars with desirable characteristics and a wide genetic base is, in turn, dependent upon breeders having access to as much genetic diversity as possible. The tea plants in China are of the broadest genetic variations in the world because of being the provenance of tea plants, long-term of allogamy and selection. Depending on these plentiful tea germplasms, China has bred more than 300 improved cultivars. A Chinese proverb says "one seed can change the whole world". Similarly, the improved tea cultivars have played critical roles in the tea industry.

The breeding objectives of tea plants have been changing from high yield to high quality then to diverse objectives such as high quality, high efficiency, high functional components contents and high tolerance to stress. Meanwhile, the breeding technique is continually advancing. A highly efficient tea breeding system, in which controlled hybridization and individual selection are the main breeding approaches combined with the molecular marker assisted selection and micropropagation techniques, has now being established and gradually developed.

1. Vegetative propagation

Vegetative propagation (VP), is the more recent of the two methods and is in the process of replacing generic propagation or seedling tea in many parts of the tea growing world (Figure 4.1). VP is asexual, as it does not use seeds which would involve the infusion of male and female gametes (germ cell or ovum). Instead, it involves taking leaf cuttings from the same "mother" bush and then planting the vegetative parts. By selecting the fish leaf, minimal damage is inflicted upon the mother bush. The rootings of these cuttings invariably have the same identical genetic constitution as the parent plant or mother bush from which they have descended and so produce identical progeny. In contradistinction to generic propagation, vegetative propagation allows for greater exactitude in breeding. The value of seed-grown plants, however, is that they exhibit greater genetic diversity. Other less common forms (they are slower to propagate) of vegetative propagation include grafting, layering and propagating root cuttings.

Figure 4.1 Vegetative propagation of tea cuttings.

扫码看彩图

2. Individual selection breeding

The selection of natural exiting variations from high quality local cultivars (jats) or open-pollinated progenies of elite cultivars is a dominant method for tea breeding. For example, six national clones, "Longjing 43", "Longjing Changye" and "Zhongcha 102" were selected from the seedlings of famous local cultivar "Longjing Qunti" and "Anhui 1", "Anhui 3" and "Anhui 7" were from "Qimen Qunti", respectively.

3. Hybridization breeding

Hybridization is one of the main methods of obtaining genetic variation, and it is an important method of breeding new varieties. For example, "Echa 1" was bred from the artificial hybridization of "Fuding Dabaicha" as female parent and "Meizhan" as male parent. Distant hybridization is a powerful method for broadening the genetic base of new varieties. Due to sterility or extremely weak fecundity, currently it could not be a routinely used method in the breeding of tea plant.

4. Mutation breeding

Mutation breeding is artificially using physical, chemical and biological factors to induce the tea plant to produce genetic variation and then breed new clones or get new valuable genetic materials for further breeding use according to the breeding objectives. The combined effects of γ-ray and chemical mutagens on biological damage to tea plant were systematically analyzed. "Zhongcha 108" is selected from the offsprings of "Longjing 43" cuttings under ^{60}Co γ-ray radiation, which is very early sprouting in the spring, high cup tea quality and resistant to disease.

The efficiency of traditional selection and breeding is very low. It will take 22-25 years to successfully breed a new clone using traditional methods. Thus, in order to shorten the breeding time and improve the breeding efficiency, lots of research projects on early-stage appraisal technique for breeding have been carried out and achieved significantly. The early-stage appraisal of tea breeding is now advancing from morphological, physiological and chemical level to more precise DNA level.

Molecular biology techniques provide polymorphic DNA based molecular markers for plant genetics and breeding. Molecular markers have distinct advantages compared to other genetic markers. Firstly, it is the direct reflection of genetic variation on DNA level, it is unaffected by the developmental stages and environmental conditions and can stably be inherited to offsprings. Secondly, some markers are co-dominant, suitable for the selection of recessive agronomical traits. Thirdly, the variation of genome is plentiful and the potential markers are almost unlimited. Presently, several molecular markers, such as RFLP, RAPD, CAPS, AFLP, ALPs, SSR, ISSR, have been developed and widely applied in the tea plant.

The significant progress of functional genomics of tea plant provide a novel and effective approach to understand the mechanisms of growth, development, differentiation, metabolism, quality and stress resistance on the whole genome level. It is possible to genetically manipulate and control the tea plant. This will bring tremendous effects to tea genetic improvement and breeding in the foreseeable future.

词汇表

genetic diversity 遗传多样性
allogamy [æˈlɒgəmɪ] 异花授粉
germplasms [ˈdʒɜːmplæzm] 种质，即亲代通过生殖细胞或体细胞传递给子代的遗传物质
micropropagation technique 微繁技术
improved cultivar 良种
vegetative propagation 无性繁殖
seedling 种苗，籽苗
mother bush 母穗
generic propagation 属内繁殖
genetic diversity 遗传多样性
grafting 嫁接
layering 压条
rooting cutting 根插
Individual selection breeding 选择育种，是从茶树群体中选择符合育种目标的类型，经过比较、鉴定和繁殖，培育出新品种的方法

> jats [dʒɑːtz] 即 Jat cultivars，群体种或有性系，用种子繁殖获得的群体
> progeny ['prɒdʒəni] n. 后代；后裔
> hybridization breeding 杂交育种，是基因型不同的配子进行交配或结合成杂种，通过对后代杂种比较、鉴定、选择、培育，获得新品种的方法
> mutation breeding 诱变育种，人为地利用物理、化学因素诱发生物产生遗传性的变异。根据育种目标对突变体进行选择和鉴定，直接或间接培育新品种的育种途径
> mutagens [mˈjuːteɪdʒenz] 诱变剂，能引起生物体遗传物质发生突然或根本改变，使其基因突变或染色体畸变达到自然水平以上的物质
> cuttings [kʌtɪŋz] 插穗，是扦插材料。用于扦插的带有成熟叶片(母叶)和健壮叶芽的小段枝条
> early-stage appraisal 早期鉴定，是指育种初期对育种材料的主要性状进行的鉴定
> molecular marker 分子标记，广义是指可遗传的并可检测的 DNA 序列或蛋白质。狭义是指能反映生物个体或种间基因组中某种差异的特异性 DNA 片段
> recessive agronomical traits 隐性的农艺性状

思考题

1. 将下面的句子译成英文。

（1）茶树育种是根据茶树遗传变异规律，通过有效创造、鉴定和筛选茶树优良遗传变异，以培育优质、高产、多抗茶树新品种的技术。

（2）大多数的茶树品种都同时具备有性(reproductive propagation)和无性繁殖的能力，因此茶树既可以用种子繁殖，也可以使用茎和根等部分营养器官进行无性繁殖。无性繁殖所得的后代性状一致，能够保护茶树良种的优良特性，繁殖系数高。常见的无性繁殖为茶树枝条扦插。

（3）采用杂交、选择和无性繁殖相结合的方法是培育茶树新品种的有效方法。

2. 将下面的句子译成中文。

（1）The development of new tea cultivars with desirable characteristics and a wide genetic base is, in turn, dependent upon breeders having access to as much genetic diversity as possible.

（2）A highly efficient tea breeding system, in which controlled hybridization and individual selection are the main breeding approaches combined with the molecular marker assisted selection and micropropagation techniques, has now being established and gradually developed.

（3）The significant progress of functional genomics of tea plant provide a novel and effective approach to understand the mechanisms of growth, development, differentiation, metabolism, quality and stress resistance on the whole genome level.

参考文献

[1] HALL N. The tea industry [M]. Philadelphia, U.S.: Woodhead Publishing, 2000：

42-248.

[2] CHEN L, ZHOU Z X, YANG Y J, Genetic improvement and breeding of tea plant (*Camellia sinensis*) in China: from individual selection to hybridization and molecular breeding [J]. Euphytica, 2007, 154: 239-248.

[3] BRAMEL P J, CHEN L. A global strategy for the conservation and use of tea genetic resources [M]. Technical Report, 2019, doi: 10.13140/RG.2.2.20411.05922.

Lesson 4 Good Agricultural Practice Management in Tea Plantation

1. GAP

The high quality and healthy food become more important and consumers have concerns about control of food production and demand information along the food chain. Good agricultural practices (GAP Figure 4.2) is based on the principals of risk prevention, risk analysis, sustainable agriculture by means of Integrated Pest Management (IPM) and Integrated Crop Management (ICM), using existing technologies for the continuous improvement of farming systems. The GAP is the utmost important for protection of consumer health. It requires ensuring safety throughout the food chain and it must be compulsory transparent not only from the table but also upstream (e.g. fertilizers, plant protection, animal feed).

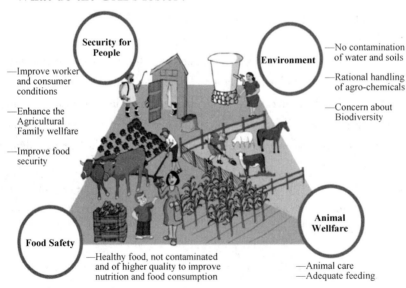

Figure 4.2 Good agricultural practice

扫码看彩图

2. GAP management

Changes in climate often affect the quantity of tea that farmers can grow. Many tea farmers are already adapting their practices to the changing climate.

One approach is agroforestry, in which tea is grown in a forest-like ecosystem with trees and shrubs instead of as a monoculture in a terraced garden. Agroforestry provides tea plants with more shade, which helps to protect them from the heat of the Sun, and also reduces the amount of moisture that tea plants lose by transpiration, protects them from frost, and helps to prevent soil erosion. If legumes are incorporated into the forest, soil can be enriched through the actions of nitrogen-fixing microorganisms that live in the roots of such plants. Agroforestry seems to improve the quality of the resulting tea, by affecting the level of phenolic compounds it contains. For tea agroforest, there's less of a drop in the level of epigallocatechin gallate and other catechins, compared to the terraced gardens.

Depending on good agricultural management in tea plantation, many tea farmers in Assam are taking steps to mitigate the effects of climate change. Soil conservation is the most common practice, with 82%–100% of plantations doing things such as: covering soil with mulch (to conserve moisture); contour farming (in which crops or drainage ditches are located along terraces that follow the slope of the land to help water better soak into the soil and prevent erosion); providing shade for tea plants; or filling in bare ground with vegetation. Some plantation managers also conserve water in ponds or behind dams for use in irrigation.

Tea gardens are set up on the cleared hill slopes where shade trees are planted in advance. Seeds are sown in the germination beds and the saplings transplanted to the garden. The garden is regularly hoed and weeded so that tea bush grows without any hindrance. Use of manures and fertilizers is a common practice in the gardens. Oil cakes and green manures are widely used. In order to increase the yield, proper dose of nitrogenous fertilizers such as ammonium sulphate should be given to soil.

The new age tea growers also have dedicated a significant part of their effort for developing a sustainable, eco-friendly model for tea production. Many growers now specialise in producing various types of handmade specialty teas which they grow in an organic or natural way.

3. GAP management adviced by Unilever for small tea farmers

The Unilever Sustainable Agriculture Initiative provides some guidance on the farming practices which will support sustainable tea production. The recommendations were set out under ten factors contributing to good practice in sustainable production.

Soil Fertility—Soil organic matter is important to provide nutrients and water. Retain tea prunings in the field. Do not take them away for fuel or any other use. Leave leaf fall and prunings from shade trees in the field. Keep the ground covered by a crop or a mulch, including prunings, whenever possible. Add manure and plant litter after pruning where organic matter is low. Consider growing Guatemala grass or legumes to rehabilitate soils for 2 years before planting tea.

Soil loss—Plant a cover crop, including beans, finger millet or maize, along the contours if land slopes significantly. Construct silt pits (short trenches in between tea rows) in newly planted areas to arrest run-off and encourage water retention. Maintain the pits until the crop covers the land. Construct drains to avoid rapid flows which cause erosion, using stones at vulnerable corners and planting grass along the sides to hold the soil. A drain across the slope will have lower flow velocity and result in less erosion than a drain directly down hill. Use tea prunings and other mulches, including litter from maize stocks, finger millet stocks, agro forestry trees and prunings from other perennial crops, to cover all bare soils that could be liable to significant erosion.

Nutrients—Use a combination of mulches and fertilizers to maintain the health of the crop. Use organic matter and compost to reduce the need for inorganic fertilizer application. Do not use ash from fires on tea fields because it will reduce the soil acidity. Dark green, fleshy and succulent shoots throughout the plucking table may indicate excess application of Nitrogen.

Pest management—The use of pesticides on mature tea should be avoided. If pesticides are used, their application must be restricted to those products recommended by the national tea research institutes and approved under national regulations. In preference to applying herbicides, manual weed management is recommended for small farms. Use cost effective mechanical methods, including the use of mulches. If weeds are a problem in mature tea, consider whether this is a result of pruning policy. A longer pruning cycle, or a taller pruning height, results in less light penetration through the tea crop and thus less weeds. If herbicide use is necessary, it must not be applied without access to appropriate herbicides. Use ultra low volume or similar technology to minimise discharge chemical levels. Spot spray with proper targeting of the weeds and do not spray areas unnecessarily.

Biodiversity—Conservation of a wide range of plant and animal species on farms and adjacent areas helps maintain the natural balance which should support future generations of farmers. Growing a range of crops will support biodiversity (and provide alternative income, or food, if the profit on tea is low). Maintain areas of land with native plant species and where wildlife can live. Native trees can be planted throughout farms without hurting the other agricultural activities. Also plant trees that can be used to control pests (for example Neem) where possible. Plant woodlots that will produce firewood, while maintaining a diversity of native species.

Product value—Manage your crop to achieve the best balance of quantity and quality of leaf, minimising costs and waste. Maintain healthy bushes and harvest them to achieve maximum yield of the leaf quality defined by the factory. Deliver the leaf to the factory or collection point quickly with minimal damage. Foreign matter, including leaves from adjacent trees and wind-breaks, must never be present in harvested leaf. Pesticide residues should never be present in harvested leaf.

Energy—Use renewable energy resources wherever possible. Larger farmer groups may consider hydro-electricity or wind-power schemes to support power needs.

Water—Water is used on farms for both irrigation and for domestic purposes. Ensure that irrigation water is applied to maximize availability to the bush with minimal run-off. Consider the use of drip irrigation rather than sprinklers for economy of water use. Consider the impact on down-stream users when extracting water from rivers. Use buildings with appropriate roofing to feed water tanks, collecting rain water for domestic use.

Social/Human capital—Good relationships with your workforce, local community, suppliers, customers and local Government are vital for long-term success of any business. Terms and conditions should be such that the turnover rate amongst permanent employees and seasonal labour is low enough to ensure that skill levels are maintained. Be a good customer, citizen and supplier – pay and supply on time and at the agreed price. Maintain good relationships with local Government and others in your local community who use land for amenity or traditional purposes.

Local economy— As a farmer you can build and sustain these communities by buying and resourcing locally. Use reliable local suppliers wherever practical. Use local employees as much as possible. Encourage employees to send their earnings to their home and family.

词汇表

Good Agricultural Practices（GAP）良好农业规范
Integrated Pest Management（IPM）害虫综合治理
Integrated Crop Management（ICM）作物综合管理
agroforestry [ˈæɡrəʊfɒrɪstri] 农林复合，农林间作
terraced garden 梯田茶园
nitrogen-fixing microorganism 固氮微生物
contour farming 等高种植，指在山坡的同等高度的地上种植农作物，坡的走向与山坡等高线一致
hoed [həud] 耕作
ammonium sulphate 硫酸铵
neem [niːm] 印度苦楝树，又称印度楝树、印度蒜楝、印度假苦楝、宁树、印度紫丁香，楝科蒜楝属植物，分布于印度、缅甸、孟加拉国、斯里兰卡、马来西亚与巴基斯坦等亚洲亚热带、热带气候地区
bio-pesticide 生物农药

思考题

1. 将下面的句子译成英文。

（1）生态茶园是运用生态学原理，因地制宜地开发和充分利用光、热、水、气、养分等自然资源，提高太阳能和生物能的利用率，有效、持续地促进茶园生态系统内物质循环和能量循环一种崭新的种植模式。

（2）有机茶园是按有机农业生产要求和有机方式种植，不使用化肥、化学农药、基因工程改良的品种。

（3）茶园中种植遮阳树不仅可以调节光照强度，还可以改善光质，增加散射光比例，从而提高茶叶品质。

2. 将下面的句子译成中文。

（1）Variations in temperature and precipitation are known to affect tea yield as well as alter the complex balance of chemicals that gives tea its flavour and potential health benefits.

（2）Agroforestry provides tea plants with more shade, which helps to protect them from the heat of the sun, and also reduces the amount of moisture that tea plants lose by transpiration, protects them from frost, and helps to prevent soil erosion.

（3）In order to keep natural organic process, neem based bio-pesticide and cow dung based fertilizers are used in the farming process.

参考文献

[1] NOWOGRODZKI A. How climate change might affect tea[J]. Nature, 2019, 566: S10-S11.

[2] UNILEVER COMPANY. Sustainable tea-Good Agricultural Practice for farmers [EB/OL]. https://www.unilever.com/Images/es_2003_sustainable_tea_good_agricultural_practice_guidelines_for-small-farmers_tcm244-409724_en.pdf. 2019-06-07.

UNIT Five   Tea Biochemistry

Lesson 1   Constituents in Fresh Tea Leaves

There exist volatile and non-volatile compounds in teas. The tea aroma is mainly dependent on the volatile compounds, while the color and the taste of tea are mainly dependent on the non-volatile compounds. In fresh tea flush there exist a wide variety of non-volatile compounds: flavan-3-ols, flavonols and flavonol glycosides, flavones, phenolic acids and depsides, amino acids, chlorophyll and other pigments, carbohydrates, organic acids, caffeine and other alkaloids, minerals, vitamins, and enzymes.

The total polyphenols in tea flush ranges from 14% to 33%. Among them, the flavan-3-ols (catechins) are the dominant group and occupy 60%—80% of the total amount of polyphenols. Four major catechins, namely (-)-epigallocatechin-3-O-gallate (EGCG), (-)-epicatechin-3-O-gallate (ECG), (-)-epigallocatechin (EGC) and (-)-epicatechin (EC), constitute around 90% of the total catechin fraction. The catechins that are water-soluble, colorless compounds contributing to astringency and bitterness in green tea.

There are at least 15 flavonol glycosides comprising mono-, di- and tri-O-glycosides based upon kaempferol, quercetin and myricetin, and various permutations of glucose, galactose, rhamnose, arabinose and rutinose, and at least 27 proanthocyanidins, including some with epiafzelechin units. In addition, some forms have a significant content of hydrolysable tannins, such as strictinin (1-O-galloyl-4, 6-O-(S)-hexahydroxydiphenoyl-β-D-glucopyranoside) and assamaicins (chalcan-flavan dimers). Relevant data on the levels of some of these compounds in green tea shoots are presented in Table 5.1.

In Chinese green teas and oolong teas, some transformations occur and lead to, for example, the production of theasinensins (flavan-3-ol dimers linked 2→2′), oolong homo-bis-flavans (linked 8→8′ or 8→6′), oolongtheanin and 8C-ascorbyl-epigallocatechin-3-O-gallate. In black tea production, the transformations are much more extensive, with some 90%

Table 5.1 Approximate composition of green tea shoots (% dry weight)

Composition	Var. *assamica*	Small-leafed var. *sinensis*	So-called hybrid of small-leafed var. *sinensis* and var. *assamica*	Large-leafed var. *sinensis*
Substances soluble in hot water				
Total polyphenols	25-30	14-23		32-33
Flavan-3-ols				
(-)-Epigallocatechin gallate	9-13	7-13	11-15	7-8
(-)-Epicatechin gallate	3-6	3-4	3-6	13-14
(-)-Epigallocatechin	3-6	2-4	4-6	1-2
(-)-Epicatechin	1-3	1-2	2-3	2-3
(+)-Catechin				4
Other flavan-3-ols	1-2			2
Flavonols and flavonol glycosides	1.5	1.5-1.7		1
Flavandiols	2-3			1
Phenolic acids and esters (despides)	5			
Caffeine	3-4	3		
Amino acids	2	4-5	2-3	
Theanine	2	2-5	2	
Simple carbohydrates (e.g. sugars)	4			
Organic acids	0.5			
Substances partially soluble in hot water				
Polysaccharides	1-2			
Starch, pectic substances	12			
Pentosans, etc.	15			
Proteins	5			
Ash				
Substances insoluble in water				
Cellulose	7			
Lignin	6			
Lipids	3			
Pigments (chlorophyll, carotenoids, etc.)	0.5			
Volatile substances	0.01-0.02			

destruction of the flavan-3-ols in orthodox processing and even greater transformation in cut-tear-curl (CTC) processing. Pu-erh tea is produced by a microbial fermentation of green tea, and it contains some novel compounds which, it has been suggested, are formed during fermentation. These include two 8-C-substituted flavan-3-ols, puerins A and B.

It is generally considered that polyphenol oxidase (PPO), which is the key enzyme in the fermentation processes that produce black teas, but there is also evidence for an important contribution from peroxidases (POD), with the essential hydrogen peroxide (H_2O_2) being generated by PPO. The primary substances for PPO are the flavan-3-ols that are converted to quinones. These quinones react further, and may be reduced back to phenols by oxidizing other phenols, such as gallic acid, flavonol glycosides and theaflavins, which are not direct substrates for PPO. Many of the transformation products are still uncharacterised. The best known are the various theaflavins and theaflavin gallates, characterized by their bicyclic undecane benzotropolone nucleus, reddish colour and solubility in ethyl acetate.

The brownish water-soluble thearubigins are the major phenolic fraction of black tea. It is assumed that the thearubigins form from the four classes of well-defined flavan-3-ol dimmers, i.e. theaflavins, theacitrins, theasinensins and theanaphthoquinones. These dimmers could be enlarged by the creation of new B-ring C-C bonds in the case of the theasinensins or by the repeat of the dimerisation where ester gallate replaces an (epi) gallocatechin unit. This oxidative cascade hypothesis is under critical scrutiny and much remains to be elucidated, but the intransigent thearubigins are gradually yielding their secrets.

Caffeine is the major purine alkaloid present in tea. The content of caffeine in tea flush is approximately 2%-5% (dry weight basis). Theobromine and theophylline are found in very small quantities. Traces of other alkaloids, e.g. xanthine, hypoxanthine and tetramethyluric acid, have also been reported.

Free amino acids constitute around 4% in tea flush. The most abundant amino acid is theanine (5-N-ethylglutamine) which is apparently unique to tea and found at a level of 2% dry weight (50% of free amino acid fraction). The precursors in the biosynthesis of theanine in tea plant have been identified as glutamic acid and ethylamine. The site of theanine biosynthesis is the root and from there it is transferred to younger leaves.

Free sugars constitute 3%-5% of the dry weights of tea flush. It consists of glucose, fructose, sucrose, raffinose and stachyose. The monosaccharides and disaccharides contribute to the sweet taste of tea infusion. The polysaccharides present in tea flush can be separated into hemicellulose, cellulose and other extractable polysaccharide fraction.

Many volatile compounds, collectively known as the aroma complex, have been detected in tea. The aroma in tea can be broadly classified into primary or secondary products. The primary products are biosynthesized by the tea plant and are present in the fresh green leaf, whilst the secondary products are produced during tea manufacture. Some of the aroma compounds, which have been identified in fresh tea leaves, are mostly alcohols including Z-

2-penten-1-ol, n-hexanol, Z-3-hexen-1-ol, E-2-hexen-1-ol, linalool plus its oxides, nerol, geraniol, benzylalcohol, 2-phenylethanol and nerolidol. The aroma complex of tea varies with the country of origin. Slight changes in climate factors can result in noticeable changes in the composition of the aroma complex. Notably, teas grown at higher altitudes tend to have higher concentrations of aroma compounds and superior flavor, as measured by the flavor index. Growing tea in a shaded environment may change the aroma composition and improves the flavour index.

Note: In Flavor index, volatile flavor compounds (VFC) can be broadly classified into two groups: VFC (I) and VFC (II). VFC (I) is dominated by C6 aldehydes and alcohols that are lipid degradation products imparting a green grassy smell to black tea and correlating negatively with price evaluation. VFC (II) constituting mainly of terpenoid compounds and products from amino acids contributes principally to the desirable sweet flowery aroma quality. The ratio of VFC (II) to VFC (I) is called the "flavor index" (Owour et al. 1989). Clones with higher levels of this index will have a better aroma quality.

词汇表

volatile ['vɒlətaɪl] *adj.* (液体等物质)易挥发的，易发散的
fresh tea leaves / fresh tea flush/ fresh tea shoots 茶鲜叶/茶新梢
flavonols and falvonol glycosides 黄酮醇及其糖苷
phenolic acids and depsides 酚酸和缩酚酸
bitterness and astringency ['bɪtənəs] [ə'strɪndʒənsɪ] 苦味和收敛性，苦涩味
kaempferol, quercetin and myricetin 山柰素，槲皮素和杨梅素
glucose, galactose, rhamnose, arabinose and rutinose 葡萄糖、半乳糖、鼠李糖、阿拉伯糖和芦丁糖
　　（芸香糖，鼠李糖与葡萄糖通过糖苷键连接形成的二糖）
permutation [ˌpɜːmjuˈteɪʃn] *n.* (一组事物可能的一种)序列，排列，排列中的任一组数字或文字
strictinin 木麻黄素
theasinensin 茶双没食子儿茶素
puerin 普洱茶素
quinone 邻醌
theaflavin 茶黄素
thearubigin 茶红素
benzotropolone 苯骈卓酚酮
scrutiny ['skruːtəni] *n.* 细阅；仔细的观察
intransigent [ɪn'trænsɪdʒənt] *adj.* 不妥协的，不让步的
alkaloid ['ælkəlɔɪd] *n.* 生物碱
theobromine [ˌθɪə'brəʊmiːn] *n.* 可可碱
theophylline [θɪə'fɪliːn] *n.* 茶叶碱
hypoxanthine [ˌhaɪpə'zænθiːn] *n.* 次黄嘌呤

> tetramethyluric acid 四甲基尿酸
> theanine [ˈθiːənɪn] *n.* 茶氨酸
> raffinose [ˈræfɪnəʊs] *n.* 蜜三糖，棉子糖
> stachyose [ˈstækɪəʊs] *n.* 水苏（四）糖
> monosaccharide and disaccharide 单糖和双糖
> polysaccharide [ˌpɒlɪˈsækəraɪd] 多糖
> linalool [lɪˈnæləəʊl], nerol [ˈnɪrəʊl], geraniol [dʒəˈreɪnɪəʊl], benzylalcohol [benzɪˈlælkəhɒl], 2-phenylethanol [fenəlˈeθənɒl], nerolidol [ˈnerɒlaɪdɒl] 芳樟醇、橙花醇、香叶醇、苯甲醇、苯乙醇、橙花叔醇

思考题

1. 将下面的句子译成英文。

（1）茶鲜叶中茶多酚总量占干物质重的14%~33%，而儿茶素占多酚总量的60%~80%。儿茶素主要有酯型儿茶素 EGCG、ECG 和非酯型儿茶素 EC、EGC。

（2）茶鲜叶的黄酮类糖苷有15种以上，主要为山奈素、槲皮素和杨梅素的单糖苷、双糖苷或三糖苷。

（3）茶红素呈棕褐色，是红茶中最主要的多酚类物质，其可能由茶黄素、theacitrin、茶双没食子儿奈素、茶萘酚醌（theanaphthoquinones）等继续氧化所形成。

2. 将下面的句子译成中文。

（1）In fresh tea flush there exist a wide variety of non-volatile compounds: flavan-3-ols, flavonols and flavonol glycosides, flavones, phenolic acids and depsides, amino acids, chlorophyll and other pigments, carbohydrates, organic acids, caffeine and other alkaloids, minerals, vitamins and enzymes.

（2）Caffeine is major purine alkaloid present in tea. The content of caffeine in tea flush is approximately 2%-5% (dry weight basis). Theobromine and theophylline are found in very small quantities.

（3）The most abundant amino acid is theanine (5-N-ethylglutamine) which is apparently unique to tea and found at a level of 2% dry weight (50% of free amino acid fraction). The precursors in the biosynthesis of theanine in tea plant have been identified as glutamic acid and ethylamine.

参考文献

[1] CLIFFORD M N, CROZIER A. Phytochemicals in teas and tisanes and their bio-availability[M]//CROZIER A, ASHIHAR H, TOMAS-BARBERAN F. Teas, cocoa and coffee-plant secondary metabolites and health. Chichester: Wiley-blackwell, 2012: 45-54.

[2] CHEN M L. Tea and Health-An overview[M]//ZHEN Y S, CHEN Z M, CHENG S

J, et al. Tea, bioactivity and therapeutic potential. London and New York: Taylor & Francis, 2002: 4-5; 110.

[3] OWUOR P O, OBANDA M A, OTHIENO C O, et al. Changes in the chemical composition and quality of black tea due to plucking standards [J]. Agricultural and Biological Chemistry, 1987, 51(22): 3383-3384.

Lesson 2 Biosynthesis of Characteristic Secondary Metabolites

1. Biosynthesis of catechins and their gallates

Flavan-3-ols (catechins) are rich in young leaves and shoots of the tea plant. Catechins, with a basic 2-phenylchromone structure, are characterized by di- or tri-hydroxyl group substitution of the B ring, the 2,3-position isomer of the C ring, and presence of a galloyl group at the 3-position of the C ring. Galloylated catechins, including EGCG and ECG, esterified with gallic acid (GA) in the 3-hydroxyl group of the flavan-3-ol units, are major catechin compounds that account for up to 76% of catechins in the tea plant.

The biosynthetic pathway of nongalloylated catechins, which include (+)-catechin (C), (-)-epicatechin (EC), (+)-gallocatechin (GC) and (-)-epigallocatechin (EGC), is well documented. Some key genes and enzymes in the pathway (Figure 5.1) include dihydroflavonol 4-reductase (DFR; EC 1.1.1.219), leucoanthocyanidin reductase (LAR; EC 1.17.1.3), anthocyanidin synthase (EC 1.14.11.19), and anthocyanidin reductase (ANR; EC 1.17.1.3).

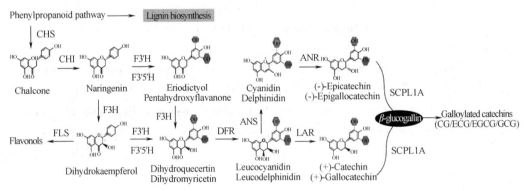

Figure 5.1 Biosynthetic pathways of the principal catechins

CHS, CHI, F3H, F3'H, F3'5'H, DFR, ANS, LAR, ANR and SCPL represent chalcone synthase, chalcone isomerase, flavanone 3-hydroxylase, flavonoid 3'-hydroxylase, flavonoid 3',5'-hydroxylase, dihydroflavonol 4-reductase, anthocyanidin synthase, leucoanthocyanidin reductase, anthocyanidin reductase, and type 1A serine carboxypeptidase-like acyltransferases, respectively.

The biosynthesis of galloylated catechins is completed in two reaction steps involving UDP-glucose: galloyl-1-O-β-D-glucosyltransferase (UGGT) and a newly discovered enzyme epicatechin: 1-O-galloyl-β-D-glucose O-galloyltransferase (ECGT) in the tea plant. β-Glucogallin, rather than GA, serves as the activated donor molecule in transacylation reactions (Figure 5.2). ECGT, an enzyme that belongs to subclade 1A of SCPL (serine carboxypeptidase-like) acyltransferases, has been shown to play critical roles in galloylation of flavan-3-ols. Transcriptome and metabolite correlation analysis showed that the tissue expression patterns for these tea-specific SCPL genes are highly correlated with the accumulation of EGCG and ECG. Particularly, it is interesting that these galloylated catechins that accumulate to high levels in young tea leaves are present mostly in the monomeric form rather than the condensed polymers proanthocyanidins (PAs), which accumulate tea in fruits, flowers, and roots. The reason is that tissues such as young buds and leaves express higher levels of LAR, ANR and SCPL than those in roots and flowers.

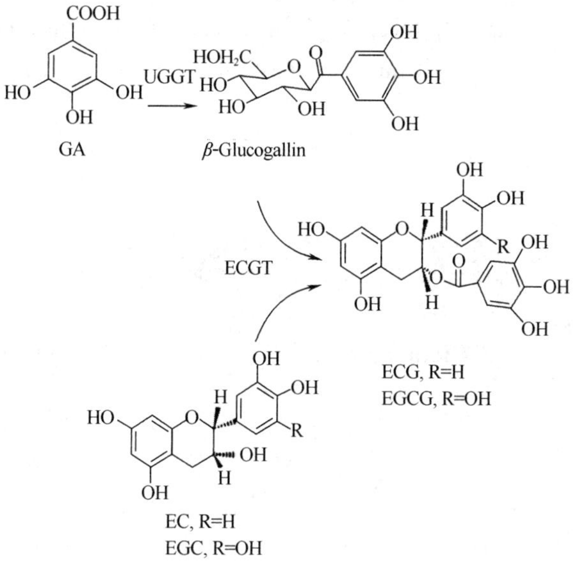

Figure 5.2 Reaction diagram of the galloylated catechin biosynthetic pathway

In the first reaction, the galloylated acyl donor β-glucogallin (βG) was biosynthesized by UGGT (UDP-glucose: galloyl-1-O-β-D-glucosyltransferase) from the substrates gallic acid (GA) and uridine diphosphate glucose (UDPG). In the second reaction, galloylated catechins ECG or EGCG were produced by ECGT (epicatechin: 1-O-galloyl-β-D-glucose-O-galloyltransferase) from the substrates βG and nongalloylated catechins EC or EGC.

2. Theanine biosynthesis

Since L-theanine has important functions in humans and plants, its distribution and biosynthesis in plants have been topics of interest. A wide variety of *Camellia* species contain

L-theanine, and it is especially abundant in *C. sinensis*, which is used to produce tea. A few *Ericales* plants and the edible mushroom *Xerocomus badius* also contain L-theanine. In *C. sinensis* plants, L-theanine is biosynthesized by L-theanine synthetase (TS, EC 6.3.1.6) from Glu and ethylamine. As L-theanine is very similar to Gln, TS is highly homologous to Gln synthetase (GS, EC 6.3.1.2). It is reported that *CsTS*1 and *GSII*-1.1 are significantly correlated with theanine content across tissues. Studies on bacteria, *C. sinensis*, and pea seeds have shown that GS can convert Glu and free ammonia to Gln, and also transfer Glu to L-theanine in the presence of ethylamine. These findings suggest that GS and TS are very similar not only in their DNA sequences, but also in their enzymatic activities.

L-Glutamic acid, a precursor of L-theanine, is present in most plants, while ethylamine, another precursor of L-theanine, specifically accumulates in *Camellia* species, especially *C. sinensis*. It is reported that *Camellia nitidissima*, *Camellia japonica*, *Zea mays*, *Arabidopsis thaliana*, and *Solanum lycopersicum* contain the enzyme/gene catalyzing the conversion of ethylamine and L-glutamic acid to L-theanine. After supplementation with [$^2H_5$] ethylamine, all these plants produce [$^2H_5$] L-theanine, which suggests that ethylamine availability is the reason for the difference in L-theanine accumulation between *C. sinensis* and other plants (Figure 5.3).

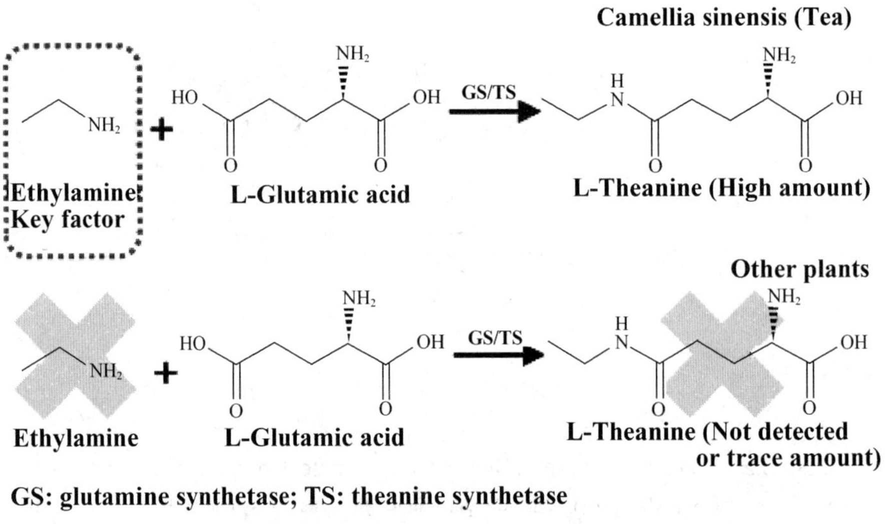

Figure 5.3 Biosynthetic pathways of L-glutamine and L-theanine in *Camellia sinensis* compared with other plants

GS, glutamine synthetase (EC 6.3.1.2); TS, L-theanine synthetase (EC 6.3.1.6).

3. Caffeine biosynthesis

As one of the most popular purine alkaloids, caffeine (1,3,7-trimethylxanthine) represents a characteristic secondary metabolite derived from purine nucleotides in many higher plants. Tea plants have relatively high concentrations of caffeine, theobromine, and theophylline. Some xanthine-based alkaloids can be synthesized *via* the synthesis of caffeine. The major caffeine biosynthetic pathway in tea plants is essentially the same as that in other purine alkaloid-accumulating plants, which begins with xanthosine and proceeds *via* three successive N-methylations of xanthosine, 7-methylxanthine, and theobromine (xanthosine (XR) → 7-methylxanthosine (7mXR) → 7-methylxanthine (7mX) → theobromine (Tb) → caffeine (Cf)) (Figure 5.4). N-methyltransferases (NMTs) were reported to catalyze the methylation steps in the major caffeine biosynthetic pathway with the methyl donor of S-adenosyl-L-methionine (SAM). Among them, caffeine synthase (CS) has ability to catalyze the final two steps (7-methylxanthine → theobromine → caffeine) and shows a bifunction of two NMTs.

In addition, previous studies with radiolabeled precursors indicated the existence of several minor pathways in tea leaves that also play a role in caffeine metabolism. Among these pathways, the 7-methylxanthine → paraxanthine → caffeine pathway and the theophylline → 3-methylxanthine → theobromine → caffeine pathway represent salvage routes for tea caffeine biosynthesis in young leave (Figure 5.4).

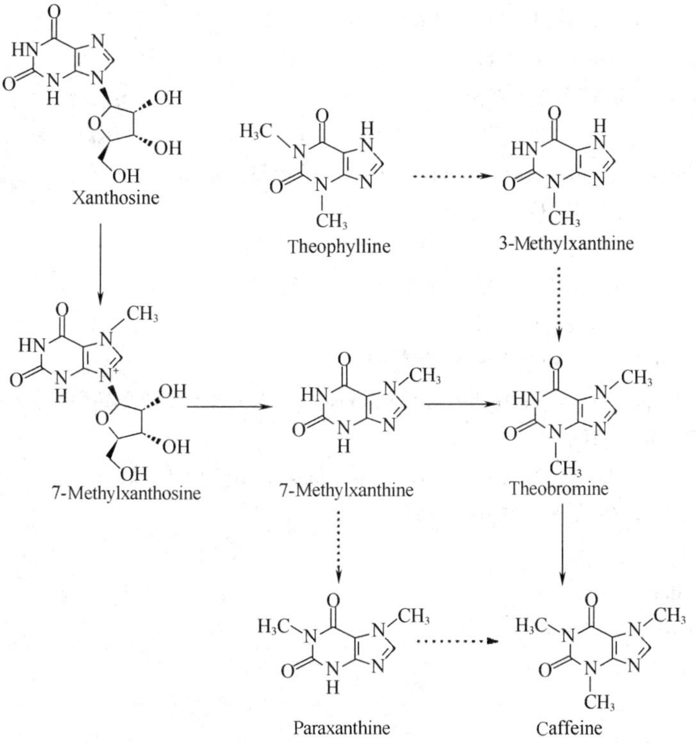

Figure 5.4 Caffeine biosynthetic pathway in tea plants (*Camellia sinensis*)

Metabolic map of the caffeine biosynthetic pathway in tea (*Camellia sinensis*) plants. The solid line indicates the major $N_7 \rightarrow N_3 \rightarrow N_1$ methylation order pathway and the dashed lines indicate the two salvaged routes for tea caffeine biosynthesis in young leaves.

词汇表

2-phenylchromone [fenɪlkˈrəʊməʊn] 2-苯基色原酮

hydroxyl group 羟基

galloyl group 没食子酰基

nongalloylated catechins 非酯型儿茶素，简单儿茶素

proanthocyanidins（PAs）原花青素

subclade 子类

phenylpropanoid pathway 苯丙烷类途径

lignin 木质素

transcriptome 转录组

naringenin 柚皮素

eriodictyol 圣草酚

leucocyanidin 无色矢车菊色素

cyanidin 矢车菊色素

chalcone synthase（CHS）查尔酮合成酶

chalcone isomerase（CHI）查尔酮异构酶

dihydroflavonol 4-reductase（DFR）二氢黄烷醇 4-还原酶

leucoanthocyanidin reductase（LAR）无色花色素还原酶

anthocyanidin synthase 花色素合成酶

anthocyanidin reductase（ANR）花色素还原酶

serine carboxypeptidase-like acyltransferase（SCPL）丝氨酸羧肽酶

flavanone 3-hydroxylase（F3H）二氢黄酮 3-羟化酶

flavonoid 3′-hydroxylase（F3′H）黄酮 3′-羟化酶

flavonoid 3′,5′-hydroxylase（F3′5′H）黄酮 3′,5′-羟化酶

flavonol synthase（FLS）黄酮醇合成酶

β-glucogallin（βG）1-O-没食子酰-β-葡萄糖

UDP-glucose: galloyl-1-O-β-D-glucosyltransferase（UGGT）尿苷二磷酸葡萄糖：没食子酰基-1-O-β-D-葡萄糖转移酶

epicatechin: 1-O-galloyl-β-D-glucose-O-galloyltransferase（ECGT）表儿茶素：1-O-没食子酰基-β-D-葡萄糖 O-没食子酰基转移酶

Ericales 杜鹃花科

Xerocomus badius 牛肝菌

Camellia nitidissima 金花茶

Camellia japonica 山茶

Zea mays 玉米

Arabidopsis thaliana 拟南芥

Solanum lycopersicum 番茄

paraxanthine [pærəˈzænθiːn] 1,7-二甲基黄嘌呤，偏黄嘌呤（拟黄嘌呤）

xanthosine（XR）[ˈzænθəsiːn] 黄嘌呤核苷

> 7-methylxanthosine 7-甲基黄嘌呤核苷
> 7-methylxanthine (7mX) 7-甲基黄嘌呤
> *N*-methyltransferases (NMTs) *N*-甲基转移酶
> *S*-adenosyl-*L*-methionine (SAM) *S*-腺苷甲硫氨酸
> caffeine synthase (CS) 咖啡碱合成酶
> salvaged route 补偿途径

思考题

1. 将下面的句子译成英文。

（1）儿茶素类化合物，属于黄烷醇类化合物。在茶树中的生物合成主要由莽草酸（shikimic acid）途径、苯丙烷途径、非酯型儿茶素合成、酯型儿茶素合成这4部分组成。

（2）咖啡碱是茶、咖啡等饮料作物中的重要生化成分。它是以黄嘌呤核苷为底物，以*S*-腺苷甲硫氨酸(SAM)为甲基供体，通过*N*-甲基化酶(NMTs)类催化的一系列甲基化反应合成的。

（3）茶氨酸是由谷氨酸和乙胺在茶氨酸合成酶的作用下，在茶树的根部合成的，在生长季节能迅速运输到叶片、枝干等地上部分。

2. 将下面的句子译成中文。

（1）The biosynthesis of galloylated catechins is completed in two reaction steps involving UGGT (UDP-glucose：galloyl-1-*O*-β-D-glucosyltransferase) and a newly discovered enzyme ECGT (epicatechin：1-*O*-galloyl-β-D-glucose-*O*-galloyltransferase) in the tea plant.

（2）As L-theanine is very similar to Gln, TS is highly homologous to Gln synthetase.

（3）The major caffeine biosynthetic pathway in tea plants begins with xanthosine and proceeds *via* three successive *N*-methylations of xanthosine, 7-methylxanthine and theobromine.

参考文献

[1] LIU Y J, GAO L P, LIU L, *et al*. Purification and characterization of a novel galloyltransferase involved in catechin galloylation in the tea plant (*Camellia sinensis*) [J]. Journal of Biological Chemistry, 2012, 287：44406-44417.

[2] WEI C L, YANG H, WANG S, *et al*. Draft genome sequence of *Camellia sinensis* var. *Sinensis* provides insights into the evolution of the tea genome and tea quality [J]. Proceedings of the National Academy of Sciences, 2018, 115(18)：E4151-E4158.

[3] CHENG S, FU X, WANG X, *et al*. Studies on the biochemical formation pathway of the amino acid L-theanine in tea (*Camellia sinensis*) and other plant [J]. Journal of

Agricultural and Food Chemistry, 2017, 65(33): 7210-7216.

[4] LI M, SUN Y, PAN S, et al. Engineering a novel biosynthetic pathway in *Escherichia coli* for the production of caffeine [J]. RSC Adv, 2017, 7: 56382-56389.

Lesson 3 Polyphenol Oxidase

1. Characteristic of polyphenol oxidases

Polyphenol oxidases (PPOs), a group of biofunctional, copper-containing oxidases having both catecholase and cresolase activities, are widely distributed within plant organisms. PPOs are referred as an oxygen and electron-transferring phenol oxidase, and responsible for browning reactions throughout catalyzing the oxidation of phenolics into quinones. In high plant, PPOs are almost ubiquitous and located in chloroplasts. Due to wounding, senescence and biotic stress, PPOs in the cytoplasm would contact vacuolar phenolic substrates in the presence of oxygen, ultimately result in enzymatic oxidation, negatively impacting on the nutritional properties and shelf life of food products. However, the enzymatic oxidation could be beneficial in some instance, as it contributes to the formation of special flavor to satisfy the demand of consumers.

2. Enzymatic oxidation in tea primary processing

One of the most important enzymes in the primary processing of fermented tea (black tea, oolong tea) is PPO, which is responsible for the conversion of original polyphenols into quality-related compounds such like theaflavins (TFs) and thearubigins (TRs). After rolling step, the cell structure of fresh leaves are damaged in order to excrete vacuoles polyphenols on the surface of the leaves. These polyphenol substrate are coupled with PPOs, giving rise to the fermentation process, inducing the following oxidative reactions and the production of flavor components.

Due to the economic importance of fermentation during tea production, PPOs have been extensively studied in relation to their capacity to oxidize phenolic compounds. Catechins, the major components of tea polyphenols, consist of (-)-epigallocatechin gallate (EGCG), (-)-epigallocatechin (EGC), (-)-epicatechin gallate (ECG) and (-)-epicatechin (EC), which are the precursors of TFs and TRs. Briefly, the conversion of tea catechins to TFs and TRs catalyzed by PPOs occurs by means of two oxidation steps (Figure 5.5). The first step is the oxidation of catechins induced by PPOs or peroxidases (PODs), leading to the formation of quinones; subsequently, the quinones are oxidized and condensed into TFs and TRs.

Comparative study on the products obtained by the POD-catalyzed and PPO-catalyzed oxidation showed that both enzymes produced the similar products. In the PPO-catalyzed

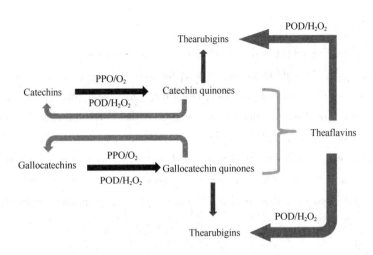

Figure 5.5 Oxidative reactions of tea catechins into theaflavins and thearubigins

oxidation, catechins were nearly expended, and the oligomers were mainly formed; as for the POD-catalyzed oxidation, a significant amount of catechins were retained, and only a small fractions of oligomers were produced, suggesting a marked higher catalytic activity of PPOs in the oxidation reactions.

3. Controlling of PPOs in post-harvest stage of plant food

Generally, the activity of PPOs activity depends on pH and temperature. The optimal pH of PPOs during black tea production ranges between 4.5 and 5.0, and the content of TFs would reach the maximum at the pH values. The catalytic activity of PPOs is also affected by temperature, since adverse temperature may result in enzyme denaturation. The optimal temperature for PPOs in most plants ranges from 25 to 50 ℃, therefore, heat treatment is applied to the inactivation of PPOs during the primary production of teas, with a temperature of 90 to 95 ℃ in tea leaf, so that the fermentation or oxidation could be ceased in a very short time, and the flavor of tea would be consolidated.

A series of chemical-based methods, including high hydrostatic pressure, high pressure carbon dioxide, ultrasound, pulsed electric fields, UV–Vis radiation, pulsed light, cold plasma and ozone processing, were introduced to control PPOs-induced quality deterioration in postharvest vegetables and fruits. Comparing with traditional thermal methods, these innovative treatments are eco-friendly for effectively reducing PPO activity, thus preventing enzymatic browning and consequently improving sensory, antioxidant and nutritional properties of plant food.

4. Conclusions

Considering the important influences of enzymatic oxidation on the quality of primary processed plant foods, the studies on the underlying mechanisms associated with PPOs activity represented a pivotal research field for food technology and plant physiology in past decades. Moreover, lots of information has been accumulated on the biochemical features of PPOs and their role in plant physiology, and the further investigation on the role of PPOs will benefit from the advances in plant genomics or functional genetics, including targeted DNA modification by genome-editing and bioinformatic approaches, facilitating better understanding of PPOs function in general, and the roles specific PPOs may play in specialized metabolism.

词汇表

polyphenol oxidase (PPO) [ˌpɔliˈfiːnɔl ˈɔksideis] 多酚氧化酶
catecholase [kəˈtɪkələs] 儿茶酚酶
cresolase [ˈkriːsəʊleɪs] n. 甲(苯)酚酶
browning reaction 褐变反应
catalyze [ˈkætəlaɪz] vt. 催化，促进
chloroplast [ˈklɒrəplɑːst] n. 叶绿体
senescence [sɪˈnesns] n. 衰老；老年期
biotic stress 生物胁迫
vacuolar [vækjʊˈəʊlə(r)] adj. 液泡的；空泡的
substrate [ˈsʌbstreɪt] n. 底物；基底
enzymatic oxidation 酶促氧化
primary Processing 初加工
POD, peroxidase [pəˈrɒksɪdeɪs] 过氧化物酶
oligomer [ˈɒlɪgəʊmə] n. 低聚物，低聚体；寡聚体
cold plasma [kəʊld ˈplæzmə] 冷等离子；低温等离子体
ozone [ˈəʊzəʊn] n. 臭氧
postharvest 采后处理；采后
enzymatic browning 酶促褐变

思考题

1. 将下面的句子译成英文。

（1）多酚氧化酶是一类广泛分布于植物体中的铜结合酶，它在细胞质中合成，通过一定的方式转运至质体内而成为具有酶活性的形式，对农产品品质形成有重要影响。

（2）多酚氧化酶是茶叶酶学中研究最多的一种酶，其对茶叶品质形成有很重要的作用，并且不同类型的茶叶生产过程中，多酚氧化酶所起的作用有所不同。

（3）茶叶加工就是采用一些技术手段钝化或激发酶的活性，使其沿着不同茶类加工所需的条件发生酶促反应而获得各类茶特有的色香味。如红茶加工过程中的发酵就是激化酶的活性，促使茶多酚物质在多酚氧化酶的催化下发生氧化聚合反应，生成茶黄素、茶红素等氧化产物，生成红茶特有的色泽和香味的过程。

2. 将下面的句子译成中文。

（1）Several studies have shown that the effect of EGCG and other tea polyphenols on key cell signaling pathways may be dependent on polyphenol-mediated reactive oxygen species.

（2）Polyphenol oxidases (PPOs) are widely distributed among organisms from bacteria to mammals, which can catalyze the oxidation of phenolics to quinones which produce dark colored melanin. PPOs play important roles in the physiological metabolism and enzymatic browning of tea plants.

（3）The quinones formed by PPOs can bind plant proteins, reducing protein digestibility and their nutritive value to herbivores. On the other hand, the oxidation of phenolic substrates by PPO is thought to be the major cause of the brown coloration of many fruits and vegetables during ripening, handling, storage and processing.

参考文献

［1］SANG S, LAMBERT J D, HO C T, et al. The chemistry and biotransformation of tea constituents［J］. Pharmacological Research, 2011, 64(2)：87-99.

［2］UYAMA H, KOBAYASHI S. Enzymatic synthesis and properties of polymers from polyphenols［J］. Advances in Polymer Science, 2006, 194(1)：51-67.

［3］TENG J, GONG Z, DENG Y, et al. Purification, characterization and enzymatic synthesis of theaflavins of polyphenol oxidase isozymes from tea leaf (*Camellia sinensis*)［J］. LWT-Food Science and Technology, 2017, 84：263-270.

［4］TARANTO F, PASQUALONE A, MANGINI G, et al. Polyphenol oxidases in crops：biochemical, physiological and genetic aspects［J］. International Journal of Molecular Sciences, 2017, 18(2)：377.

［5］QUEIROZ C, LOPES M L M, FIALHO E, et al. Polyphenol oxidase：characteristics and mechanisms of browning control［J］. Food Reviews International, 2008, 24(4)：361-375.

［6］TAKEMOTO M, TAKEMOTO H. Synthesis of theaflavins and their Functions［J］. Molecules, 2018, 23(4)：918.

［7］TINELLO F, LANTE A. Recent advances in controlling polyphenol oxidase activity of fruit and vegetable products［J］. Innovative Food Science & Emerging Technologies, 2018, 50：73-83.

［8］SULLIVAN M L. Beyond brown：polyphenol oxidases as enzymes of plant specialized metabolism［J］. Front Plant Sci, 2015, 5：783.

[9] BUTT M S, IMRAN A, SHARIF M K, et al. Black tea polyphenols: a mechanistic treatise [J]. Crit Rev Food Sci Nutr, 2014. 54(8): 1002-1011.

[10] DRYNAN J W, CLIFFORD M N, OBUCHOWICZ J, et al. The chemistry of low molecular weight black tea polyphenols [J]. Natural Product Reports, 2010, 27(3): 417-462.

Lesson 4 Glycosidase

1. Characteristic of glycosidase in tea plant

Glycosidase (O-glycoside hydroases, EC 3.2.1.X), are a widespread category of enzymes of significant biochemical, which involve in the glycosides hydrolysis (Ⅰ), transglycosylation (Ⅱ), and condensation (Ⅲ) reactions (Figure 5.6). Reactions Ⅰ and Ⅱ are forward reactions, while reaction Ⅲ is reverse reaction in hydrolysis, accompanying by the synthesis of oligosaccharide.

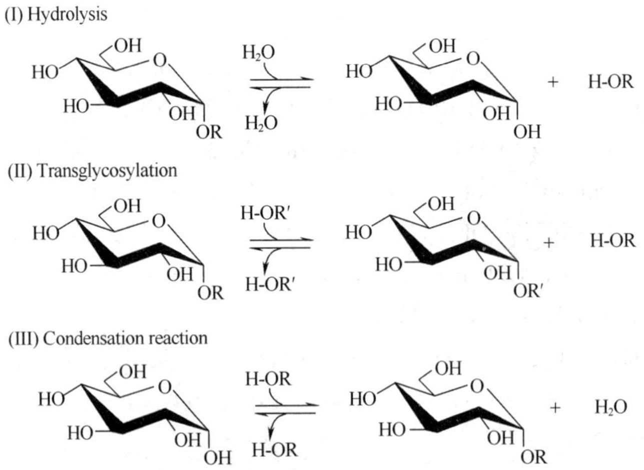

Figure 5.6 Simplified reaction models of glycosidases (R=aglycone)

In tea plant, most of the volatile compounds exist in cells as glycosidic precursors, since glycosylated volatile compounds are more water soluble and stable than their free form. The glycosylated volatiles found in tea leaves are mainly in form of β-primeverosides or glucosides, and a large portion of these aroma precursors distribute in young leaves. Glycosidases, primarily catalyze the hydrolysis of the glycosidic bonds into oligo – or polysaccharides, liberating free volatiles, and forming the characteristic aroma of various teas.

Two enzymes, namely β-primeverosidase and β-glucosidase, have been purified and identified from the fresh leaves of tea plant, which play vital roles in the hydrolysis of glycosidically-bound volatile compounds. Although the protein sequence is 50% to 60% identical to that of β-glucosidases, β-primeverosidase was not able to hydrolyze β-D-glucopyranoside. Additionally, β-primeverosidase demonstrates very high selectivity to primeveroside, and specifically hydrolyzed the β-glycosidic bond between the disaccharide and the aglycons. The β-primeverosidase was highly accumulated in young leaves, and decreased as the leaf maturity, consistent with the distribution of the glycosidically bound aroma compounds. Interestingly, the β-primeverosidase locates in cell walls and the cavity areas among cells, whereas the glycosidically-bound aroma compounds are present in vacuoles, suggesting the enzyme and the substrate would not couple in living tea plant, unless the cell structure is damaged and exposed to the enzymes.

2. Changes in glycosidase activities during the tea manufacturing process

The studies on glycosidases have been carried out during the manufacturing processes of black and oolong teas, and it was found that floral fragrance in black and oolong teas were probably derived from the enzymatic hydrolysis of volatile compounds during the manufacturing processes.

In freshly plucked tea leaves, primeverosides were 3-fold higher than glucosides; after stepping into manufacturing process, the contents of primeverosides decreased greatly, whereas the abundance of glucosides were not remarkably changed, implying primeverosides are hydrolyzed during the manufacturing. Moreover, the glycosidases activities are peaked at withering stage, but significantly decreased after rolling, suggesting that glycoside hydrolysis mainly takes place during the rolling stage of black tea production. Therefore, glycosidases-catalyzed glycosides hydrolysis plays a key role during the manufacturing of black tea. By contrast, there are no obvious glycosidic bound volatile compounds reductions observed within the manufacturing process of oolong tea. The levels of most glycosides rise after the solar-withering, reaching peak values in final stage of oolong tea production, although the underlying mechanism of the increasing of glycosides is still unclear. The expression of glucosidase genes could be induced by insect infestation, manufacturing stresses and mechanical injury. It's also reported that ultraviolet B irradiation on tea leaves could stimulate the genes expression of β-primeverosidase and β-glucosidase, contributing to the release of volatiles from the leaves.

3. Conclusions

Many volatile compounds in fresh tea leaves are in the forms of non-volatile precursors or glycosides, which are liberated by glycosidase during tea processing. The precursors and glycosidase enzymes are varied depending on tea plant varieties and processing methods. The numerous category of volatiles giving off from living or processed tea leaves indicate that there are still much more questions to figure out, such as, most investigations on glycosidically-bound tea volatiles have focused on hydrolysis reaction, while the transglycosylation and condensation effects of glycosidases have received much less attention. Additional molecular studies are needed to obtain a comprehensive understanding on the role glycosidases play in the metabolism process of tea plant.

词汇表

glycosidase [glaɪˈkəʊsɪdeɪs] n. 糖苷酶
aglycone [æɡliːˈkəʊn] 糖苷配基
glycosidic precursor 糖苷前体
primeveroside [praɪmiːvəˈrəʊsaɪd] 樱草糖苷
primeverosidase [praɪmiːvəˈrəʊsɪdeɪz] 樱草苷酶
glucoside [ˈɡlukəˌsaɪd] n. 葡(萄)糖苷；糖苷
glucosidase [gluːˈkəʊsɪdeɪs] n. 葡(萄)糖苷酶
aroma precursor 香气前体

思考题

1. 将下面的句子译成英文。

（1）香气是茶叶感官品质的重要评价指标之一。糖苷类香气前体物质水解后通常表现出不同的香型风味，是茶叶主要的香气前体。

（2）糖苷在植物中分布极广，是糖和糖的衍生物等与另一非糖物质(苷元)失去水分子，通过苷键缩合而成的化合物。

（3）在茶叶生长及加工过程中，因虫食、害虫侵染、失水或机械损伤等，导致茶叶细胞区室化功能破坏，使得茶叶中的糖苷态香气前体物质，经糖苷酶分解释放出挥发性物质。

2. 将下面的句子译成中文。

（1）In tea leaves, the glycosidically-bound volatile compounds require the action of glycosidases to liberate free volatiles. A β-primeverosidase and a β-glucosidase have been purified from fresh leaves of tea plant.

（2）On the basis of the foregoing, buds and the first leaves were found to contain large amounts of aroma precursors and high glycosidase activity, indicating that high quality tea are reasonably made from young tea leaves from aroma formation mechanistic points of view.

（3）Various aroma glycosides have been identified in fresh tea leaves. The chemical structure of most glycosides was shown to be β-primeveroside, suggesting that a common biosynthetic machinery exists for the conjugation of β-primeveroside to aroma volatiles.

参考文献

[1] CHIBA S. A Historical perspective for the catalytic reaction mechanism of

glycosidase; so As to bring about breakthrough in confusing situation [J]. Journal of the Agricultural Chemical Society of Japan, 2012, 76(2): 215-231.

[2] WANG D, KURASAWA E, YAMAGUCHI Y, et al. Analysis of glycosidically bound aroma precursors in tea leaves. 2. Changes in glycoside contents and glycosidase activities in tea leaves during the black tea manufacturing process [J]. Journal of Agricultural & Food Chemistry, 2001, 49(4): 1900-1903.

[3] SPECIALE G, THOMPSON A J, DAVIES G J, et al. Dissecting conformational contributions to glycosidase catalysis and inhibition [J]. Current Opinion in Structural Biology, 2014, 28: 1-13.

[4] RYE C S, WITHERS S G. Glycosidase mechanisms [J]. Current Opinion in Chemical Biology, 2000, 4(5): 573-580.

[5] VOCADLO D J, DAVIES G J. Mechanistic insights into glycosidase chemistry [J]. Current Opinion in Chemical Biology, 2008, 12(5): 539-555.

[6] ZHENG X Q, LI Q S, XIANG L P, et al. Recent advances in volatiles of teas [J]. Molecules, 2016, 21(3): 338.

[7] YANG Z, BALDERMANN S, WATANABE N. Recent studies of the volatile compounds in tea [J]. Food Research International, 2013, 53(2): 585-599.

[8] OHGAMI S, ONO E, HORIKAWA M, et al. Volatile glycosylation in tea plants: sequential glycosylations for the biosynthesis of aroma beta-primeverosides are catalyzed by two *Camellia sinensis* glycosyltransferases [J]. Plant Physiology, 2015, 168(2): 464-477.

Lesson 5 Quantification of Catechins, Caffeine and Theanine by HPLC

1. Application of high performance liquid chromatography in teas

Catechins are classified as monomers of flavonoids, which are known as the major biological active compounds from teas. Caffeine is an alkaloid compound of xanthine, which is naturally present in tea and coffee. There is no difference regarding the caffeine content between green and black teas, which suggests that caffeine is very stable during the manufacturing process. L-theanine is the most abundant amino acid in tea plant, accounting for over 50% of the total free amino acid content, 1%-2% dry weight of fresh leaf.

The enormous variety of analytical methods has been developed to quantify catechins, caffeine and theanine contents in teas. Reverse-phase high performance liquid chromatography (HPLC) followed by UV or electrochemical or mass determination is the most widely used approach for analysis of secondary metabolites in tea. The first application of HPLC for the detection of tea constituents was reported in 1976 by Hoefler and Coggon, in which the content of five tea catechins (C, EC, ECG, EGC and EGCG) in green tea infusion were quantified. Subsequently, improvements for the measurement have been implemented by Goto and co-workers in 1996. The same year, thermospray-LC-MS analysis was introduced to quantify the caffeine and theanine from tea. The focus of recently analytical HPLC method studies are mainly on optimization of the analysis time per sample, the conditions used to extract, and characterize components found in infusions or biological matrix, as well as high-throughput detection.

2. Operational procedures of quantification of catechins and caffeine

For catechins and caffeine quantification, the sample solution could be prepared as described in latest Chinese National Standard (GB/T 8313—2018). Briefly, a sample of tea powder (0.2 g) was extracted with 5 mL 70% (V/V) methanol at 70 ℃ for 10 min with stirring (every 5 min), followed by centrifuge at 1400 g for 10 min under room temperature. After centrifuge, the supernatant was transferred into a new 10 mL volumetric flask. Then we repeated the extraction and combined all the supernatant together which made up to 10 mL with 70% (V/V) methanol. The extracts were filtered with 0.22 μm filter, then subjected to HPLC analysis.

The HPLC conditions are as follows: C_{18} column (particle size 5 μm; column size 250 mm × 4.6 mm), the column temperature was maintained at 28 ℃. The gradient was from solvent A (9% acetonitrile, 2% acetic acid, 0.2% EDTA, 88.8% water) to solvent B (80% acetonitrile, 2% acetic acid, 0.2% EDTA, 17.8% water), and the linear gradient condition of the mobile phase is 0→10 min, 100% A; 10→25 min, 100%→68% A; 25→35 min, 68% A; 35→40 min, 68%→45% A; 40→45 min, 45% A; 45→50 min, 45%→100% A; and 50→60 min, 100% A. The flow rate is 1 mL/min, and the detection wavelength is set at 278 nm. The peak retention times and samples areas are determined by HPLC and the quantitative analysis is conducted according to the external standard method.

3. Operational procedures of quantification of theanine

A tea sample of 1.0 g was extracted with 300 mL of deionized water for 20 min at 100 ℃. The extract was vacuum filtered, then transferred to a new 500 mL flask and diluted with water to a constant volume and mixed. Subsequently, the extract was filtered through a 0.22 μm filter and transferred into a centrifuge tube. Finally, pre-column derivatization was performed by using AccQ-Fluor Reagent Kit according to the manufacturer's specifications. The test solution was subjected to HPLC for the analysis of amino acids.

The HPLC conditions are as follows: AccQ.Tag reversed-phase HPLC column (particle size 4 μm; column size 3.9 mm × 150 mm) is used, and the column temperature is maintained at 25 ℃. The mobile phases comprise AccQ.Tag Eluent A (A), acetonitrile (B), and MilliQ water (C). The following are the gradient programs used for separation of the amino acids in tea samples: 100% A at 0 min, turned linearly to 91% A, 5% B and 4% C at 17 min, then change into 80% A, 17% B and 3% C at 24 min, 68% A, 20% B and 12% C at 32 min, and the last for 2 min, then go to 60% B and 40% C at 35-37 min, and return to 100% A at 38 min, then add to column wash and stabilization from 38 to 45 min. The flow rate is 1 mL/min, and the detection wavelength is set at 395 nm. The peak retention times and samples areas are determined by HPLC, and quantitative analysis is carried out according to the external standard method.

4. Conclusions

In recent published literatures, tea was investigated for chemical characteristic using LC-MS metabolomic analysis and chromatographic methods, and the contents of catechins, caffeine, and theanine were used as chemical descriptors to identify the categories of tea samples. The most important advances of HPLC approach in near future, would be high-throughput detection with improvements in separation efficiency and sensitivity. The application of ultra-high performance liquid chromatography (UHPLC) to the analysis of tea components has shown clear benefits in terms of analysis time, resolving power, and solvent consumption. HPLC coupled with CoulArray electrochemical detector (ECD) is the most commonly used analytical method to measure tea polyphenols in biological fluids, with 100-fold more sensitive than UV detector. LC with fluorescence detection and chemiluminescence (CL) detection is also a sensitive method that could determine trace components of tea sample *in vivo*. HPLC with mass detection (LC-MS), especially with tandem mass spectrometry (MS/MS), is the most powerful technique for structure identification of trace quantities of tea constituents and the metabolites of tea constituents. Electrospray ionization (ESI) and atmospheric pressure chemical ionization (APCI) are the two most commonly used ionization methods to analyze tea constituents.

Analytical methods play a very important role in the identification of the tea constituents. The aforementioned HPLC methods developed in recent years provide convenient tools for quick identification and quantification of biochemical active compounds in teas, with the progress in instrumental techniques and the field of biochemistry, more effective methods without the tedious process including bioassay-guided extract pre-fractionation, isolation and purification of the active compounds would be the future development trends of chromatographic technique.

词汇表

high performance liquid chromatography 高效液相色谱

reverse-phase 反相

quantify [ˈkwɒntɪfaɪ] vt. 定量；确定……的数量

thermospray [θɜːməspˈreɪ] 热喷（法），热喷雾

LC-MS，Liquid Chromatography-Mass Spectrometry 液质联用

biological matrix [ˌbaɪəˈlɒdʒɪkl ˈmeɪtrɪks] 生物基质

high-throughput [haɪ ˈθruːpʊt] 高通量；高通量的

Chinese National Standard 中国国家标准

centrifuge [ˈsentrɪfjuːdʒ] v. 离心 n. 离心机

supernatant [ˌsjuːpəˈneɪtənt] 上清液；上清

volumetric flask [vɔljuˈmetrɪk flɑːsk] （容）量瓶

filter [ˈfɪltə(r)] n.&v. 过滤

gradient [ˈgreɪdiənt] n. 梯度；adj. 倾斜的

solvent [ˈsɒlvənt] n. 溶剂

acetonitrile [æsɪtəʊˈnaɪtrɪl] n. 乙腈

EDTA abbr. ethylene diamine tetraacetic acid 乙二胺四乙酸

mobile phase 流动相

flow rate 流速

detection wavelength 检测波长

retention time（RT）保留时间

external standard method 外标法

deionized water [diːˈaɪənaɪzd] 去离子水

vacuum filter 真空过滤；减压抽滤

centrifuge tube 离心管

pre-column derivatization 柱前衍生化

AccQ Fluor Reagent Kit 氨基酸衍生用试剂盒

MilliQ Water 水是 Millipore 公司的主要产品，其生成的水属于超纯水，即其电阻率为 $18.2 M\Omega \cdot cm$

metabolomic analysis [metæbəʊˈlɒmɪk] 代谢组学分析

chemical descriptor 化学变量

separation efficiency and sensitivity 分离效率和灵敏度

CoulArray electrochemical detector（ECD）库拉雷电化学检测器

chemiluminescence [ˌkemɪˌljuːmɪˈnesəns] n. 化学发光

trace component 痕量成分

Electrospray ionization（ESI）电喷雾电离

atmospheric pressure chemical ionization（APCI）大气压化学电离

bioassay 生物鉴定；生物测定

pre-fractionation 预分离

思考题

1. 将下面的句子译成英文。

（1）本研究利用超高效液相色谱-串联质谱建立了一种可以同时检测茶叶8种儿茶素类物质的分析方法，应用本方法可以检测出茶叶样品中儿茶素、表儿茶素、表没食子儿茶素、表儿茶素没食子酸酯、表没食子儿茶素没食子酸酯等物质的含量。

（2）茶叶中儿茶素及咖啡碱可以通过丙酮、乙醇、甲醇、乙腈或水等为溶剂在一定温度与时间参数条件下提取出来。准确地定量这两类成分，为衡量茶叶品质、指导茶叶加工等领域的重要技术保障。

（3）茶氨酸是茶叶中特有及最主要的游离氨基酸，它对茶汤的滋味和香气均有良好的作用，可以缓解茶汤的苦涩味，增强茶汤鲜甜滋味，是绿茶鲜爽滋味的主要来源。

2. 将下面的句子译成中文。

（1）Tea is abundant in chemical components that differ in both their contents and component ratios, which not only results in multiple dimensions of tea quality but also leads to some difficulties in the qualitative analysis of tea, especially when tea chemical information far exceeds the subjective analysis of the taster. Thus, quantitative indicators need to be utilized to identify tea.

（2）Among various analytical methods, high-performance liquid chromatography (HPLC) has been widely used by the main exporting and importing countries to identify chemical compositions in tea products, providing high accuracy and sensitivity.

（3）The aim of this study is to develop an accurate and quantitative method based on the chemical compositions of teas. Numerous representative samples covering six tea categories corresponding to different processing methods are collected worldwide. Catechins, caffeine and theanine are employed as distinct variables to develop internationally recognized tea models and provide references to aid the international tea trade and consumers.

参考文献

［1］EL-SHAHAWI M S, HAMZA A, BAHAFFI S O, et al. Analysis of some selected catechins and caffeine in green tea by high performance liquid chromatography［J］. Food Chemistry, 2012, 134(4): 2268-2275.

［2］SANG S, LAMBERT J D, HO C T, et al. The chemistry and biotransformation of tea constituents［J］. Pharmacological Research, 2011, 64(2): 87-99.

［3］CSUPOR D, BOROS K, JEDLINSZKI N, et al. A validated RP-HPLC-DAD method for the determination of L-theanine in tea［J］. Food Analytical Methods, 2014, 7(3): 591-596.

［4］GUILLARME D, CASETTA C, BICCHI C, et al. High throughput qualitative analysis of polyphenols in tea samples by ultra-high pressure liquid chromatography coupled to

UV and mass spectrometry detectors [J]. Journal of Chromatography A, 2010, 1217(44): 6882-6890.

[5] PENG W B, TAN J L, HUANG D D, et al. On-line HPLC with biochemical detection for screening bioactive compounds in complex matrixes [J]. Chromatographia, 2015, 78(23-24): 1443-1457.

[6] KALILI K M, VILLIERS A. Recent developments in the HPLC separation of phenolic compounds [J]. Journal of Separation Science, 2011, 34(8): 854-876.

[7] TAO W, ZHOU Z, ZHAO B, et al. Simultaneous determination of eight catechins and four theaflavins in green, black and oolong tea using new HPLC-MS-MS method [J]. Journal of Pharmaceutical and Biomedical Analysis, 2016, 131: 140-145.

[8] NING J, LI D, LUO X, et al. Stepwise identification of six tea (*Camellia sinensis* (L.)) categories based on catechins, caffeine, and theanine contents combined with Fisher discriminant analysis [J]. Food Analytical Methods, 2016, 9(11): 3242-3250.

UNIT Six Tea Processing and Machinery

Lesson 1 Tea Classification

1. Basic tea classification system

Tea is the most popular and healthy beverage, all derived from the processed young tea shoots of the tea plant *Camellia sinensis*. Normally, tea can be classified into two categories: basic tea, reprocessed or further-processed tea products. There are a vast number of different teas including flavoured tea, teabags, decaffeinated tea, and tea or flavoured tea with other food ingredients, which belongs to the re-processed or further-processed form from its original material "tea", belonging to the basic tea types.

So far, most of the knowledge surrounding tea classification is generally used Chinese tea classification system. In this system, the teas were classified into six basic tea types recognized as green, yellow, white, oolong (blue), black and dark tea, which based on the processing method and the oxidation (also called fermentation) extent of tea polyphenols especially catechins in tea fresh leave after being plucked. Actually, the processing method is to accurately control the duration and manner of oxidation. The oxidation can be caused by endogenous oxidase, microbe oxidase, damp heat and dry heat. In China, this system has developed into National Standard GB/T 30766-2014 "classification of tea", implemented on October 26th, 2014 issued by General Administration of Quality Supervision, Inspection and Quarantine of the People's Republic of China (AQSIQ) [*Note*: *AQSIQ was restructured and merged into State Administration for Market Regulation (SAMR) in 2018.*] and Standardization Administration of P.R.C (SAC).

All unique tea processing technique are to determine the oxidation ways and extents in tea classification system as follows.

Fixation, in green tea processing inactivates the oxidase activity and halts the oxidation. This process is accomplished by moderately heating tea leaves, therefore deactivating their oxidative enzymes and removing undesirable scents in the leaves, without damaging the flavour of the tea.

Maceration (rolling) and subsequent fermentation, in black tea processing promote and accelerate the oxidation by cell endogenous oxidases. The maceration (Rotorvane/Lawrie Tea Processer/Cut Torn Crushed (CTC)/Rollers) breaks down the structures inside and outside of the leaf cells and allows, from the co-mingling of oxidative enzymes with various substrates, which allows the beginning of oxidation. Subsequently, the macerated leaf is held in warm, humid air for up to 6 hours depending on maceration degree of tea leaves.

Bruising of leaf edges, in oolong tea controls the oxidation at the edge of mature tea shoots.

Long-time withering, of tea shoots in white tea controls the oxidation slowly with the cells broke naturally when mositrue evaporation.

Yellowing, of tea shoots in yellow tea is a thermal oxidation under warm atmosphere by piling or packed after shoots fixed (inactivated/deactivated the oxidases).

Post-fermentation, in dark tea promotes the oxidation by the microorganism excreted oxidases under wet-piling after shoots fixed (inactivated the oxidases).

Apart from generally known Chinese tea classification, there are other tea classification methods like Europian method, Japanese method.

Europe— Tea in Europe is categorized as black tea, oolong tea, green tea and further types including non-fermentation/aeration (e.g. white tea), part-fermentation/aeration (e.g. yellow tea) and post-fermentation/aeration (e.g. Pu-erh tea). This category is similar to Chinese tea classification.

Japan— Tea classification is similar to Chinese method as shown in Figure 6.1.

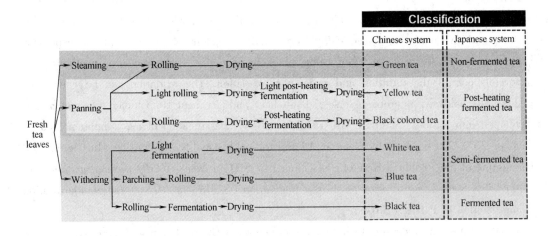

Figure 6.1 Japanese classification of tea in the world

2. Basic six tea types could be further-processed or re-processed

Any of the above six tea types can be further processed into miscellaneous tea products as flavored, scented, blended, ground, aged, decaffeinated tea, or tea extracts and so on. In the world, the loose tea types are preferred in Asian countries, while teabag, tea beverage, tea ingredient supplements are the main consumed tea products in western countries.

词汇表

tea processing and machinery 茶叶加工与机械
tea classification 茶叶分类
reprocessed or further-processed 再加工或深加工
flavoured tea [ˈfleɪvəd tiː] 加香茶；香味茶
teabag 袋泡茶；茶包
decaffeinated tea [ˌdiːˈkæfɪneɪtɪd tiː] 脱咖啡碱茶
food ingredient [fuːd ɪnˈɡriːdiənt] 食品成分；食品配料；食品配料成分
green, yellow, white, oolong, black and dark tea 绿茶、黄茶、白茶、青(乌龙)茶、红茶和黑茶
endogenous oxidase [enˈdɒdʒənəs] 内源氧化酶
microbe oxidase [ˈmaɪkrəʊb] 微生物氧化酶
General Administration of Quality Supervision, Inspection and Quarantine (AQSIQ) 国家质量监督检
　　验检疫总局(现为国家市场监督管理总局"State Administration for Market Regulation, SAMR")
Standardization Administration of P.R.C (SAC) 国家标准化管理委员会
fixation 杀青
maceration 破碎
Rotorvane 洛托凡碎茶机(转子叶片)
Lawrie Tea Processer(LTP) 锤击式揉切机
Cut Tear Crush (CTC) 切碎—撕裂—粉碎
roller 揉捻机
bruising [ˈbruːzɪŋ] v. (使)出现伤痕；擦伤
bruising of leaf edges 摇青
moisture evaporation 水分散失
yellowing 闷黄
post-fermentation 后发酵
aeration [eəˈreɪʃn] n. 充(通、换、透)气
Pu-erh tea(或 Pu'er tea, Puer tea) 普洱茶
aged 陈化
steaming 蒸青
panning 炒青

思考题

1. 将下面的句子译成英文。

(1) 根据制作方法和茶多酚氧化(发酵)程度的不同,可分为六大类:绿茶(不发酵)、白茶(微发酵)、黄茶(轻发酵)、青茶(乌龙茶,半发酵)、黑茶(后发酵)和红茶(全发酵)。

(2) 以各种毛茶或精制茶叶再加工形成再加工茶,包括花茶、紧压茶、萃取茶、药用保健茶、茶食品、含茶饮料等。

(3) 深加工茶是指用茶的鲜叶、成品茶叶为原料,或是用茶叶、茶厂的废次品、下脚料为原料,利用相应的加工技术和手段生产出的茶制品。茶制品可能是以茶为主体的,也可能是以其他物质为主体的。

2. 将下面的句子译成中文。

(1) Actually, the processing method is to accurately control the duration and manner of oxidation. The oxidation can be caused by endogenous oxidase, microbe oxidase, damp heat and dry heat.

(2) The maceration (Rotorvane/Lawrie Tea Processer/Cut Torn Crushed/Rollers) breaks down the structures inside and outside of the leaf cells and allows, from the co-mingling of oxidative enzymes with various substrates, which allows the beginning of oxidation.

(3) Yellowing of tea shoots in yellow tea is a thermal oxidation under warm atmosphere after shoots fixed (inactivated/deactivated the oxidases).

参考文献

[1] Tea & Herbal Infusion Europe. Compendium of Guidelines for Tea (*Camellia sinensis*) [EB/OL]. http://www.thie-online.eu/fileadmin/inhalte/Publications/Tea/2018-08-20_Compendium_of_Guidelines_for_Tea_ISSUE_5.pdf. 2019-06-07.

[2] Classification of tea in the world [EB/OL]. http://www.o-cha.net/english/cup/pdf/4.pdf. 2019-06-07.

Lesson 2　Green Tea

1. Traditional hand processing

Leaf that is to be manufactured into green tea is first picked over carefully for the removal of any remaining leaf stalks and foreign materials. At the earliest possible moment after plucking, the leaf is placed in a much deeper *kuo* and is given a preliminary firing that amounts to a thorough fixation. The green tea *kuo* is set in the top of a waist-high, brick stove, and about five inches below the level of its surface. The *kuo*, itself, is about sixteen inches in diameter and ten inches deep, making a total depth of fifteen inches from the top of the stove. For pan-firing the leaf, the *kuo* is made very nearly red hot with a wood fire, and into it about half a pound of leaves are thrown by tea maker. Then he stirs the leaves rapidly and they produce a crackling sound and a great quantity of steam. The operator frequently raises the leaves a little above the top of the stove, and shakes them over the palms of his hands to separate them and allow the steam to penetrate. At length, after two or three final brisk turns about the *kuo*, the operator collects the leaf together in a heap, and, with a single deft motion, sweeps it into a basket held in readiness by another workman.

From the *kuo* the pan-fried leaf goes at once to a table covered with matting for hand-rolling (Figure 6.2). After the leaves have been rolled into a ball, they are shaken apart and then twisted between the palms of the hands; the right hand passing over the left with a slight degree of pressure as the hand advances, and relaxing again as it is returned. This twists the leaves regularly and in the same direction. After rolling, the leaf is spread out in sieves and allowed to cool for a short time; then it again goes to the *kou* for first drying. This time the fire is considerably diminished, and charcoal, instead of wood, is used for fuel, in order to avoid any smoke; but the pan is still kept so hot that the finger cannot be borne upon it for more than an instant. Close attention is given to regulating the heat; a fireman being constantly employed in this duty.

Figure 6.2 Hand-rolling

扫码看彩图

The operator alternately stirs and rolls the leaf with his hands and twists in the same manner as at the rolling table. This process is continued until the leaves have lost so much of their sap as to produce no more steam. After that, the leaves are merely stirred about in the pan until they are comparatively dry, turn a dark olive color, and, after being removed from the *kuo* and sifted, are once more put back and given a final drying. In the final-drying a peculiar change of coloring takes place in the leaf. It assumes a bluish tint, which is a distinguishing characteristic of this sort of tea after it has reached the correct degree of dryness. The total processing occupy about ten hours.

The product is called *mao ch'a* (primary tea), when not sorted and graded on the spot, is packed in chests, and sold by the farmers and peasants to collectors, who in turn sell it to the *tea hongs* (tea shops) in the larger cities. There it passes through a succession of sifting processes, and is winnowed by being tossed about on bamboo trays or by various forms of pedal-driven fanning mills. It is then separated into commercial grades, such as Gunpowder, Imperial, Young Hyson and Hyson.

2. Processing

Although different green teas may be produced by different processing techniques, the general green tea processing is achieved as follows:

Fresh tea leaves plucked → fixed (inactivating enzyme, *called shaqing in Chinese*) → rolled/shaped → dried (or fired).

It is worthy to note that different kinds of green tea have different requirements for plucked young tea shoots. The common young tea shoots required are one bud and two or three leaves, while those for famous high-quality green tea are usually one bud and one leaf, even just one bud. In Japan, some high-quality green teas such as gyokuro and matcha are produced from the fresh tea leaves grown under 90% shading for 2 weeks and 40%-50% shading for 1-2 weeks before being plucked, respectively. The quality of fresh tea leaves plays a key role in the characteristics of green tea.

In the harvesting season, the tender tea shoots are plucked by hand or a mechanical tea plucker and delivered to tea factories immediately. In order to prevent fermentation (the oxidation of catechins) and formation of the characteristic "green leaf, green liquor" of green tea, the leaves are immediately fixed to inactivate the endogenous enzymes by steam or heat with pan frying, roasting or baking. However, on the heavy harvesting days, the leaves are thinly spread out indoors in a shallow basket in the racks, or some ventilating beds like bamboo mesh mats, trays, troughs, or spreading machines, and wait for fixation. The waiting time should be as short as possible to avoid oxidation, which turns the green leaf and stem to red. The tea fixation apparatus include a tea steaming machine, wood-fired pan, electrical pan, rotary heated drum, microwave and far infrared. When fixed leaves become soft and flaccid, they are conveyed to be rolled, and then the rolled tea masses are loosened by a roll breaker or ball breaker and moderately cooled. Subsequently, the leaves are fired (dried) by charcoal-fired baking baskets, electrical heaters, coal heaters, liquid petroleum gas heaters, or natural gas heaters. The moisture content of the final product (crude tea or primary tea) should be less than 6%.

According to the fixation ways, green tea is subdivided into steamed green tea, pan-fried green tea and sun-dried green tea. Steamed green tea is mainly produced and consumed in Japan, while pan-fried green tea is mainly produced in China and exported to the world.

Crude green teas are usually refined by the wholesale tea dealers. The refining process generally includes sifting, cutting, grading, refiring, polishing, blending, and packing. Through refining, the stalks, dust, and impurities are removed. The final tea has a uniform size and standardized appearance, such as Chunmee, Sowmee, and Gunpowder. Chunmee and Sowmee are renowned for their curve shape, like the human eyebrow, while Gunpowder is featured with tiny, tight, round balls like gunpowder.

词汇表

kuo [译音] *n.* 锅(wok)
steam *n.* 水蒸气
deft motion 灵巧的动作
pan-fried 锅炒的
hand-rolling 手揉
spread out 摊开
sap [sæp] *n.* 液，汁
tint [tɪnt] *n.* 色调；淡色彩
tea hongs 茶行
sift 筛选；挑选；精选
winnow ['wɪnəʊ] *v.* 风选，簸扬
toss [tɒs] *v.* (轻轻地)颠簸，摇摆，挥动
pedal-driven 踏板驱动
Gunpowder tea 珠茶
Imperial tea 御茶
Young Hyson 雨茶(另：Chunmee 珍眉；Hyson 贡熙；Sowmee 秀眉)
sorting ['sɔːtɪŋ] *n.* 分类拣选；整理
shape *v.* 做形
Gyokuro *n.* (日)玉露(茶)
mechanical tea plucker 机械采茶机
shallow basket ['ʃæləʊ 'bɑːskɪt] 浅口匾篮
ventilating beds 通风床
trough [trɒf] *n.* 槽
spreading machine 摊青机
steaming machine 蒸青机
rotary heated drum 可加热滚筒
far infrared 远红外

> soft and flaccid ['flæsɪd] 柔软无光泽的
> roll/ball breaker 解块机
> charcoal-fired baking basket 炭火烘笼
> heater 加热器
> moisture content 含水量(水分含量)
> steamed green tea 蒸青绿茶
> pan-fried green tea 炒青绿茶
> refiring 复火
> polishing 车色
> blending 拼配
> dust tea 茶末,意为切得像粉一般细的茶叶(注意与现在将茶叶加工成粉的茶粉 tea powder、超微茶粉 superfine tea powder、速溶茶粉 instant tea 之间的区别)
> impurity [ɪmˈpjʊərəti] n. 杂质

思考题

1. 将下面的句子译成英文。

(1) 杀青是绿茶加工的关键工序,目的是防止茶多酚的氧化而产生红变。

(2) 绿茶按其干燥和杀青方法的不同,一般分为炒青、烘青、晒青和蒸青绿茶。

(3) 炒青绿茶的精制,主要是通过筛分、切断、风选、拣剔等作业,分成各种大小、长短、粗细、轻重不同的筛号茶,再经过复火、车色、清风割末等工序,最后将各种筛号茶分等级、匀堆拼配为成品茶。

2. 将下面的句子译成中文。

(1) From the *kuo* the pan-fried leaf goes at once to a table covered with matting for hand-rolling. After the leaves have been rolled into a ball, they are shaken apart and then twisted between the palms of the hands; the right hand passing over the left with a slight degree of pressure as the hand advances, and relaxing again as it is returned. This twists the leaves regularly and in the same direction.

(2) In order to prevent fermentation (the oxidation of catechins) and formation of the characteristic "green leaf, green liquor" of green tea, the leaves are immediately fixed to inactivate the endogenous enzymes by steam or heat with pan frying, roasting or baking.

(3) Through refining, the stalks, dust, and impurities are removed. The final tea has a uniform size and standardized appearance, such as Chunmee, Sowmee and Gunpowder.

参考文献

[1] WAN X C, LI D X, ZHANG Z Z. Green tea and black tea: Manufacturing and Consumption [M]//HO C T, LIN J K, SHAHIDI F. Tea and tea products: Chemistry and

health-promoting properties. Floride, U.S.: CRC Press, 2009: 1-8.

[2] UKERS W H. All about tea(II) [M]. New York, U.S.: The Tea and Coffee Trade Journal Company, 1935.

Lesson 3 Black Tea

1. Origin of black tea

Black tea came out in the early Ming Dynasty. A legend says that Zhu Yuanzhang went to a battle, arrived at the foot of Wuyi Mountain. People in the local area were so scared that they escaped. The troops stationed in a farmer's home and slept on the fresh tea leaves. When the troops left and the famer went back home on the second day, the leaves turn to red because of a night's fermentation. The famer cried, but thought it a pity to throw away those leaves. He processed a new kind of tea with the red leaves and sold it at the local county-Tongmuguan. Unbelievably, the tea was sold at a good price and was processed from then on. Today, it is called the Lapsang Souchong Tea, the ancestor of black tea.

Black tea is a fully fermented tea. It has red liquid and red leaves after infusion. So, in China it is called *Hongcha* (red tea, 红茶) in reference to the color of the infused liquid or to the red edges of the oxidized leaves, as opposed to the black color of the main body of the final dried tea leaves. Usually, the aroma of Congou black tea is sweet with a kind of ripe fruits. Sometimes, it has a lovely and pleasing scent of flowers too. The taste of black tea is mellow and sweet.

2. Processing

The most widely used form, black tea, is produced by a variety of methods. However, the general outline procedure is common as: fresh tea leaves plucked → withered → rolled → fermented → dried.

Among those four steps, the fermentation process is crucial to the quality of the final black tea product, which is predominated with the oxidation of catechins and production of oxidation reaction products.

The typical plucked tea flush for black tea processing is one bud and two leaves. On arrival at the factory, they are spread out on large trays, racks or mats, troughs, or a machine and are left to wither by natural air current under sunshine or indoor controlled ventilation/aeration with the aid of warm-air fans. The moisture in the leaves evaporates and the leaves become limp and flaccid. Subsequently, the leaves are processed by orthodox roller or rotorvane, or CTC (crushing, tearing and curling) machine, or LTP (Lawrie tea processor) machine. Most of the black tea in India, Sri Lanka, and Kenya is manufactured using the CTC process, while that in

China is processed principally by traditional orthodox rollers. The objective of the rolling is to break the leaf cells and release the oxidases, including polyphenol oxidase and peroxidase, and initiate the process of catechin oxidation with oxygen in the air. Importantly, CTC can be used to handle efficiently large volumes of tea leaves, rapidly rupturing withered tea leaves to small particles and forcing out most of cell sap, which leads to sufficient fermentation.

After rolling, the broken tea leaves are transferred to the fermentation room and laid out thinly on trays, in troughs, or on the floor at a little warm (25-35℃), high humidity (>95%) atmosphere for fermentation. The fermentation time ranges from a half hour to 6 hours, depending on the variety of tea plants, the age of tea leaves, the particle size of broken tea leaves, and the fermentation condition. Among these factors, the rupturing technique plays a key role. Generally, tea leaves macerated by CTC machine need a short time, from 30 to 60 minutes, while tea leaves ruptured by orthodox roller take a long time, from 2 to 6 hours. In this process, the broken tea leaves set to fully oxidize, which starts during rolling. Due to the oxidation, green leaves gently turn to golden russet color and the greenish leaf note turns to a fresh or floral aroma.

As the optimum fermentation is achieved, the leaf mass is dried or fired to inactivate the enzymes and halt the fermentation. Continuous driers are usually used, in which hot air is generated by electrical heater or coal furnace. In this process, the leaf turns dark brown or black, the aroma changes to floral, and the moisture is reduced to less than 6%.

The crude black teas produced in the world are mainly congou (Gongfu) black tea and CTC black tea, which are processed by orthodox rolling and CTC machine, respectively. Apart from those two, there is still a minor productivity Souchong black tea produced in the Wuyi (Bohea) mountain area in China. It is said that souchong black tea was created in the middle of the fifteenth century, and lapsang souchong (Zhengshan Xiaozhong) is known to be the origin of black tea. The processing of lapsang souchong is similar to that of congou black tea, except that the fermented dhool (refers to the tea leaf during fermentation, noted for its coppery color) is fired at 200 ℃ for several minutes and rerolled before final drying. In addition, souchong black tea can be further processed by the absorbance of the scents released from the burning pine branches to smoked souchong black tea, which is known for its smoked flavor.

Similar to crude green tea refining, crude black teas are refined through sifting, cutting, grading, blending, refiring, and packing. Through the refining, the stalks, fibers, and impurities in crude tea are removed. The fine teas of CTC are graded to four varieties as whole leaf grades, brokens, fannings and dusts. Moreover, each variety is subdivided into several categories as summarized in the following list.

Whole leaf grade—Special finest tippy golden flowery orange pekoe (SFTGFOP), finest tippy golden flowery orange pekoe (FTGFOP), tippy golden flowery orange pekoe (TGFOP), golden flowery orange pekoe (GFOP), flowery orange pekoe (FOP), flowery pekoe (FP), pekoe souchong (PS), orange pekoe (OP), pekoe (P), souchong (S).

Brokens—Flowery broken orange pekoe (FBOP), golden flowery broken orange pekoe

(GFBOP), golden broken orange pekoe (GBOP), tippy golden broken orange pekoe (TGBOP), tippy golden flowery broken orange pekoe (TGFBOP), broken orange pekoe (BOP), broken pekoe (BP), broken pekoe souchong (BPS).

Fannings—Orange fannings (OF), broken orange pekoe fannings (BOPF), pekoe fannings (PF), broken mixed fannings (BMF).

Dusts—Pekoe dust (PD), red dust (RD), fine dust (FD), golden dust (GD), super red dust (SRD), super fine dust (SFD).

词汇表

Lapsang Souchong Tea 正山小种
Congou black tea 工夫红茶
mellow and sweet 醇甜
one bud and two leaves 一芽二叶
warm-air 暖风
orthodox ['ɔːθədɒks] adj. 传统的；正统的
CTC 红碎茶
Bohea 武夷；武夷茶；武夷山
lapsang souchong 正山小种
dhool 发酵叶，茶坯
reroll 复揉
russet ['rʌsɪt] adj. 赤褐色的 n. 赤褐色
whole leaf grades, brokens, fannings and dusts 叶茶、碎茶、片茶和末茶
pekoe 白毫(茶)

思考题

1. 将下面的句子译成英文。

（1）内源性儿茶素和氧化酶在红茶加工过程中从细胞中释放出来，在滚切过程中结合在一起，氧化形成较高分子质量的茶黄素和茶红素，从而形成红茶的红汤红叶。

（2）经茶发酵适度一般以叶色基本变为红黄色，青气消失，花果香显现为标准。发酵温度一般控制在 22-30℃，具体需根据鲜叶原料，萎凋、揉捻程度等因素综合确定。

（3）国际红茶分为叶、碎、片、末四个等级，是基于鲜叶等级和加工完成后的形状尺寸进行分级，形成不同的等级规格。

2. 将下面的句子译成中文。

（1）Subsequently, the leaves are processed by orthodox roller or rotorvane, or CTC (crushing, tearing and curling) machine, or LTP (Lawrie tea processor) machine. Most of the black tea in India, Sri Lanka and Kenya is manufactured using the CTC process, while that in China is processed principally by traditional orthodox rollers.

(2) Among these factors, the rupturing technique plays a key role. Generally, tea leaves macerated by CTC machine need a short time, from 30 to 60 minutes, while tea leaves ruptured by orthodox roller take a long time, from 2 to 3 hours.

(3) In addition, souchong black tea can be further processed by the absorbance of the scents released from the burning pine branches to smoked souchong black tea, which is known for its smoked flavor.

参考文献

[1] WAN X C, LI D X, ZHANG Z Z. Green tea and black tea: manufacturing and consumption [M]//HO C T, LIN J K, SHAHIDI F. Tea and tea products: Chemistry and health-promoting properties. Floride, U.S.: CRC Press, 2009: 1-8.

[2] UKERS W H. All about tea(II) [M]. New York, U.S.: The Tea and Coffee Trade Journal Company, 1935.

Lesson 4 Oolong Tea (Blue Tea)

Oolong (blue) tea, a semi-fermented tea, with its own unique aroma and taste, has become a popular consumption as indicated by the increasing production. Varying degrees of oxidation made oolong tea (OT) varieties are endless possibilities. OT leaves are showing green in the center and red at the edge, and it is commonly called green leaves with red edging. Unlike either black tea or green tea, oolong tea has an excellent characteristic with the combination of the freshness of green tea and the fragrance of black tea. It is produced by a special process called green leaf shaking (Yaoqing, Figure 6.3) and green leaf cooling (Liangqing). In this process, the moderately withering green tea leaves are bruised at the edges by hand or mechanical shaking and vibrating. The leaf appearance of oolong tea is featured with the reddish edges and green centers. Oolong tea is produced in China, particularly in Fujian, Guangdong, and Taiwan, and is currently popular in China and Southeastern Asia. It has a good function in helping body building and dieting.

The name OT came into English language from its Chinese name (Wu Long, 乌龙) meaning "black dragon tea". The name originally had nothing to do with dragons, rather it was named after Wu Long who discovered OT by accident. Wu Long was a tea grower and hunter. One day, he was distracted by a deer after collecting a good load of tea. He stopped tea picking and tried to slay the beast. By the time he remembered to return to the tea, it had already started to oxidize. He did not want to let tea go to waste, therefore he finished tea process. Surprisingly, this tea was mellow and aromatic, unlike any tea he had tasted before.

Figure 6.3　Oolong tea manipulation machine

扫码看彩图

1. Processing

First, to produce good OT, the tea leaves must be collected at a particular time. Second, the leaves must be processed with proper methods. Processing is the most important factor in determining the quality of OT. The processes in OT can be generally classified into seven steps, sunning and withering, fermenting, panning, rolling, firing, final-firing and packing. The full process is depicted in Figure 6.4.

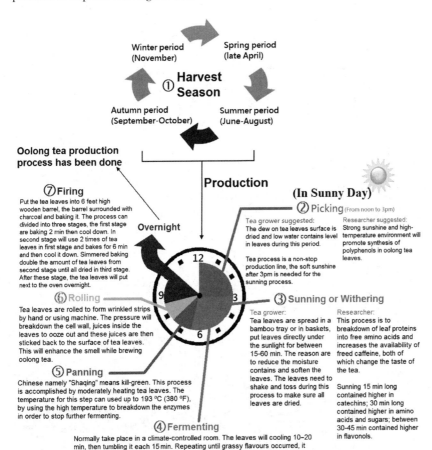

Figure 6.4　Processing of oolong tea

扫码看彩图

2. Varieties

There are six of the most common varieties of OT in China (Table 6.1). These are Tie Guan Yin (TGY), Da Hong Pao (DHP), Phoenix Dan Cong Tea (PDC), Dongfang Meiren (DM), Pou Chong Oolong (PCO), and Dong Ding (DD). The degree of oxidation and oxidation process (Figure 6.5) both directly affect the characteristics of the final product. The degree of oxidation can be up to 90%, depending on the variety and production style.

Table 6.1 General information regarding the six most common varieties of oolong tea produced in China

Varieties	Translated English name or common name	Distribution	Degree of oxidation		comments
			Taiwan official recom-mended	Tea grower recom-mended	
Pou Chong Oolong (PCG)	Pou Chong Oolong	Taipei in Taiwan	8%–12%	Less than 15%	① A lightly oxidated, usually not baked; ② With slight green tea flavors; ③ With floral and melon fragrances and rich, mild taste.
Dong Ding (DD)	Frozen Summit	Nantou in Taiwan	15%–30%	20%–35%	①Ball rolling technique seals in the sweet floral notes and flavors; ② Sweet taste accompanied by floral notes; tea is golden and amber color after brew.
Tie Guan Yin (TGY)	Iron Buddha Rolled Oolong	Anxi in south Fujian	15%–30%	20%–50%	① One of the most widespread oolong tea; ② Can be collected 4 times a year, the best is from spring tea period; ③ The fragrance is lasting after brewed and taste sweet.
Dongfang Meiren (DM)	Oriental Beauty; White Tip Oolong	Hsinchu in Taiwan	50%–60%	60%–75%	①A heavily fermented oolong tea; ② Always pick tea leaves which attacked by *Jacobiasca formosana*; ③Bright-reddish orange tea liquor produces sweet taste and has natural fruity aromas.
Phoenix Dan Cong Tea (PDC)	Single Bush	Phoenix Mountain in Guangdong	ND	70%–80%	①Extremely high prices in the top grades; ② The tea color is clear bright yellow and has unique natural fragrance.

续表

Varieties	Translated English name or common name	Distri-bution	Degree of oxidation		comments
			Taiwan official recom-mended	Tea grower recom-mended	
Da Hong Pao (DHP)	Big Red Robe Wuyi Rock Tea	Wuyi in north Fujian	ND	85%–90%	①A nearly fully-oxidized oolong; ②Always grown among the rocks; ③According to the leaf edge become red color after processing, look like wearing a big red robe; ④After brewing, it is very fragrant, with the aroma of orchid.

ND = no data.

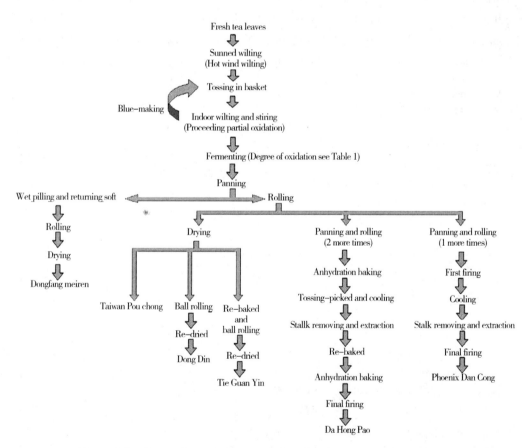

Figure 6.5 Comparison of the preparation for six varieties of oolong tea

3. The quality of oolong tea

The attributes of oolong tea can be partially correlated with the levels of certain flavor compounds. The aromas of oolong tea are due to linalool, geraniol, 2-phenylethanol, benzyl alcohol, methyl salicylate, linalool oxides, (Z)-3-hexenol, and others. The characteristic aromas of oolong tea are nerolidol, jasmine lactone, methyl jasmonate, and indole. The formation of these volatile compounds mainly involves hydrolysis of primeverosides and glucosides, the transformation of higher fatty acids into alcohols and aldehydes, the decompoisition of carotenoid into aromatic compounds, oxidation reactions of catechins and amino acids, and transformation of amino acids.

Furthermore, the flavor of oolong tea infusion is related to a combination of various compounds, such as catechins and flavones glycosides (bitterness), amino acids (freshness), soluble sugar (sweetness), theaflavins (briskness) and thearubigin (mellowness). In addition, the color of oolong tea infusion can be determined by the major combinations of flavone, catechins, theaflavins and thearubigins. According to Roberts, the appearance of the leaves in oolong tea can be related to the proportion of chlorophyll, xanthine and carotene.

词汇表

oolong(blue) tea [ˈuːlɒŋ] 乌龙茶(青茶)
oxidation [ˌɑksɪˈdeʃən] *n.* 氧化
green leaves with red edging 绿叶红镶边
green leaf shaking 摇青
green leaf cooling 凉青
orchid [ˈɔːkɪd] *n.* 兰花
wilting 萎凋
blue-making 做青
wet-piling 湿闷
panning 锅炒
stalk removing 去梗
linalool [lɪˈnæləʊl] 芳樟醇
geraniol [dʒəˈreɪnɪəʊl] 香叶醇
methyl salicylate [ˈmeθil sæˈlisileit] 水杨酸甲酯
(Z)-3-hexenol [ˈheksɪnɒl] 顺-3-己烯醇
carotenoid [kəˈrɒtənɔɪd] *n.* 类胡萝卜素

思考题

1. 将下面的句子译成英文。

（1）乌龙茶做青过程是叶细胞在机械力的作用下不断摩擦损伤，形成以多酚类化合物酶性氧化为主导的化学变化，以及其他物质的转化与累积的过程，逐步形成花香馥郁、滋味醇厚的内质和绿叶红边的叶底。

（2）乌龙茶又称青茶，属于半发酵茶，以其特有的做青工艺形成其独特的品质。

（3）乌龙茶春茶香高味厚，品质最佳；秋茶香气高锐而味薄，品质次之；夏茶香低味较苦涩，品质最差。

2. 将下面的句子译成中文。

（1）Oolong（blue）teais produced by a special process called green leaf shaking（Yaoqing）and green leaf cooling（Liangqing）. In this process, the moderately withering green tea leaves are bruised at the edges by hand or mechanical shaking and vibrating.

（2）The flavor of oolong tea infusion is related to a combination of various compounds such as catechins and flavones glycosides（bitterness）, amino acids（freshness）, soluble sugar（sweetness）, theaflavins（briskness）and thearubigin（mellowness）.

（3）Oolong teas have been processed from new shoots with one bud and four or five leaves, which help to form the unique aroma of elegant floral, ripe fruit flavor.

参考文献

［1］NG K W, CAO Z J, CHEN H B, *et al*. Oolong tea: a critical review of processing methods, chemical composition, health effects and risk［J］. Critical Reviews in Food Science and Nutrition, 2019, 58(17):2957-2980.

［2］CHEN Y L, DUAN J, JIANG Y M, *et al*. Production, quality and biological effects of oolong tea（*Camellia sinensis*）［J］. Food Reviews International, 2010, 27:1-15.

Lesson 5　White Tea

1. Processing history of white tea

The history of white tea, which originated in China, is complex and controversial, due to the lack of adequate citation as much of the knowledge about tea has been passed on verbally.

Perhaps first produced during the Tang Dynasty（618-907 AD）, "white tea"（perhaps albino tea）was the preferred tea of the Chinese royal court. It became revered during China's Song Dynasty（960-1279 AD）. By 1,200 AD, the immature silver white buds and leaves were immediately steamed, dried and ground into a powder, which was then whisked in bowls to make the delicate tea during the Song Tea Ceremony. As ruling dynasties changed, so did the production of white teas. It was virtually unknown outside of China, and was produced quite differently to the way it is today.

2. Modern process of white tea

The modern process can be traced to the late 1700s, when loose teas from a mixed-variety tea bush were steeped for tea production. By 1885, selected types of tea bushes were used to create various types of white teas. Technically, white tea may be considered to be the oldest form of tea. Other processing techniques, developed later, led to the production of other types of tea.

1) Material selection and picking standards

White tea comes from the same plant (*Camellia sinensis*) as green and black teas. It is comprised of the delicate buds and/or young leaves and the descriptive term "white" stems from the high proportion of silvery buds harvested from the plants to produce the tea. There are many grades of white tea, depending on the different plucking criteria; these are Silver Needle with White Hair, White Peony, Gongmei, and Soumei. Silver Needle is made of the pure bud, while White Peony is made of the bud and one or two leaves of Dabaicha or Shuixian varieties. Gongmei is made from the Qingzhong variety, and Shoumei is manufactured from fresh leaves without the bud.

White tea is harvested when the buds on the trees are still covered with fine white hair and the leaves have not yet fully opened. After plucked, fresh tea leaves are sunshine withered and dried. White tea is picked then withered and dried, without fermentation, rolling or roasting. White tea is kept relatively fresher than those of green or black tea. With a lighter and sweeter taste after processing, white tea is closer to fresh tea leaves than green tea, and is far different from the other varieties of tea.

The ideal time and weather for harvesting is a sunny morning when the sun is high enough to have dried any remaining moisture on the buds. No harvesting is performed on rainy days or when frosts are on the ground.

2) Withering and drying process technology

Compared with other kinds of tea, white tea is subjected to the simplest processing procedure, including only prolonged withering and drying processes without any steps of enzyme deactivation or fermentation. This particular manufacturing process confers on white tea unique flavors with a slightly sweet and umami taste as well as a fresh and green odor.

Both long withering and drying contribute to white tea aroma characteristics. In the prolonged withering process, catalyzed by endogenous enzymes of fresh leaves, a slight fermentation (oxidation) occurs, which produce the unique aroma and taste of white tea. Aroma-related key genes participated in the regulation of white tea aroma formation. Therefore, scientists systematically investigate the changes in volatile compounds during white tea processing via GC × GC – TOF/MS technology. A total of 172 volatile compounds were identified, mainly consisting of endogenous volatiles, which displayed various change trends during the withering period. The amount of most free amino acids increased significantly during the withering period, with corresponding amino acid derived volatiles (AADVs) presenting similar trends in variation. Moreover, the variation in glycosidically bound volatiles (GBVs) and key genes showed that they actively participated in the regulation of white tea aroma formation during the withering period. It demonstrated that prolonged withering constitutes a key step in white tea aroma formation through modulation of the volatile compound profile.

Nowadays, the energy – efficient sun withering room and a novel variable – frequency continuous withering machine were used for white tea leaf withering to optimize the processing condition, which aims to substantially improve white tea quality, while reduce the energy budget and labor cost by realizing fully automation of white tea leaf withering. From a published report, sun withering room for white tea in Fuding city should be facing 5 degree south by west, with the roof angle of 22°. Variable-frequency continuous withering machine can realize automatic alternation of sun withering and natural withering. The daily variation of total solar radiation is in the range from 34 to 695 W/m^2, and accordingly the daily variation of temperature is from 17.8 ℃ to 32.7 ℃, both of which comply with the requirement of white tea withering condition. It demonstrates that the quality of white tea withered in the facility is much better than that withered by traditional natural withering.

In addition, the drying procedure also played an important role in the formation of white tea aroma. It has showed that the relative contents of alcohols and esters decreased with the increase of drying temperature, while ketones, aldehydes, hydrocarbons and furans increased. The critical temperature points of the aroma profile change were 60 ℃, 100 ℃ and 120 ℃. Aromatic hydrocarbons, C10 monoterpenes and esters were the representational VFC in white tea at different temperatures. Aldehydes and ketones had additive effect on the aroma of white tea; the positive effect of aldehydes on the aroma was greater than that of ketones. However, alcohols had a negative effect on the aroma of white tea. Esters were the key reasons of why the aroma of white tea changed from unpleasure to pleasure. The increasing contents of the C10–C11 aromatic hydrocarbons and C10 monoterpene led to a further aggravation of scorch odour and the bitter smell.

3. Summary

Minimal processing not only protects the delicate, light and slightly sweet flavor of white tea, but also enables the retention of high levels of phytochemicals, which are believed to be responsible for the tea's health benefits. Because of the minimal processing, manufacturing of white tea requires the least amount of time and labor. Therefore the higher cost of white tea is primarily due to the selection and acquisition of raw materials.

词汇表

albino tea [ælˈbiːnəʊ tiː] 白化茶
revere [rɪˈvɪə(r)] v. 尊敬；崇敬
Silver Needle with White Hair, White Peony, Gongmei, Soumei 白毫银针；白牡丹；贡眉；寿眉
prolonged withering 长时萎凋
glycosidically bound volatiles (GBVs) 糖苷键合态挥发物
variable-frequency continuous withering machine 变频连续萎凋机
aromatic hydrocarbon 芳香烃
ketone [ˈkiːtəʊn] n. 酮
aldehyde [ˈældɪhaɪd] n. 醛，乙醛
hydrocarbon [ˌhaɪdrəˈkɑːbən] n. <化>碳氢化合物，烃
furan [ˈfjʊərən] n. 呋喃
monoterpene [mɒnəˈtɜːpiːn] n. 单萜
scorch [skɔːtʃ] n. 烧焦处，焦痕
phytochemical [faɪtəʊˈkemɪkəl] adj. 植物化学的

思考题

1. 将下面的句子译成英文。

（1）中国《茶经》记述白茶属于微发酵茶，萎凋、晒干或烘干是白茶加工的两道关键工艺。

（2）晒干或烘干作为白茶加工的最后一道工序，通过高温促使茶叶中挥发性成分和非挥发性成分发生一系列复杂而深刻的变化，提高茶叶品质，形成特有的香气。

（3）对萎凋叶进行摊堆也是新工艺白茶加工的一大特点，通过摊堆轻微发酵，使萎凋叶软化，为轻揉捻创造条件。

2. 将下面的句子译成中文。

（1）Four major varieties of white tea from China are available: Silver Needle (Baihaoyinzhen), White Peony (Baimudan), Long Life Eyebrow (Shoumei), and Tribute Eyebrow (Gongmei). Another white tea comes from the Darjeeling region of India: Darjeeling White.

（2）Silver Needle is produced only from the delicate top buds covered with tiny silvery/white fuzz without stems or leaves. The buds used in Silver Needle tea are picked for a period of less than one month between late March and early April.

（3）Considered a lower-grade tea, "eyebrow" teas are named for the long, thin, crescent-shaped leaves used in production. Aged after oxidation, the Tribute Eyebrow offers a dark brew with an earthy taste.

参考文献

[1] MAO J T. White tea: The plants, processing, manufacturing, and potential health benefits[M]// PREEDY V R, Tea in health and disease prevention. San Diego, U.S.: Academic Press, 2013:33-40.

[2] JIANG H Y. White tea: Its manufacture, chemistry, and health effects[M]// HO C T, LIN J K, SHAHIDI F. Tea and tea products: Chemistry and health-promoting properties. Floride, U.S.: CRC Press, 2009:17-28.

[3] ZHAO F, QIU X, YE N, et al. Hydrophilic interaction liquid chromatography coupled with quadrupole-orbitrap ultra high resolution mass spectrometry to quantitate nucleobases, nucleosides, and nucleotides during white tea withering process[J]. Food Chemistry, 2018, 266:343-349.

[4] CHEN Q, ZHU Y, DAI W, et al. Aroma formation and dynamic changes during white tea processing[J]. Food Chemistry, 2019, 274: 915-924.

[5] CHEN J B, JIN X Y, HAO Z L, et al. Research of sunlight withering room and its withering effect on white tea [J]. Transactions of the Chinese Society of Agricultural Engineering, 2012, 28(19):171-177.

[6] QIAO X Y, WU H L, CHEN D. Effects of drying temperatures on the volatile flavor compounds in white tea[J]. Modern Food Science and Technology, 2017, 33(11):171-179.

Lesson 6 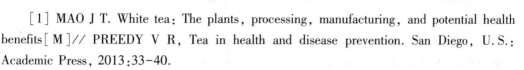 Dark Tea

Dark tea is one of the six type teas in China with nearly a thousand-year history. It is also the most special tea widely produced in China with various styles. In recent years, the dark tea industry has grown particularly rapidly, and its production/sales surpassed black tea, ranking second in the Chinese tea industry. The development of the millennium has laid the foundation of unique processing for dark tea, and the participation of microbes makes it a real fermented tea. In general, the processing technology of dark tea includes fixing, rolling, piling and drying processes, of which piling contributes greatly to its distinctive characteristics. Furthermore, the differences in tea leaves as well as processing technology have formed the unique flavor and taste of dark tea in different producing areas.

1. Introduction

Dark tea is mostly consumed by domestic market while the sales of export also have increased in recent years. In 2018, the production of dark tea is 318,900 tons, accounting for 12.2% of total Chinese tea production and 14% of total Chinese tea sales.

In the manufacturing of dark tea, particular biochemical substances formed under the reaction of microbial extracellular enzymes. The transformation compositions play a crucial role in lipid metabolism and gastrointestinal regulation. Because of this, the consumption of dark tea is more popular in frontier nationalities.

According to its different produced regions and processing technology, dark tea can be categorized into Sichuan dark tea (south road bian-xiao tea, west road Bian xiao tea), Hubei dark tea (Qingzhuan brick tea, etc.), Hunan dark tea (Dark brick tea, Fuzhuan brick tea, etc.), Yunnan dark tea (Ripe Pu-erh tea, etc.), Guangxi dark tea (Liubao tea, etc.), etc. Among all teas, Hunan dark tea, Yunnan dark tea and Guangxi dark tea have the higher sales than others.

2. Processing

Fresh tea leaves—Dark tea fresh leaves are more mature than those for ordinary green and black tea (Table 6.2). Generally, the tea leaves are plucked at summer, autumn or the end of spring tea season. Meanwhile, the tea leaves normally plucked mechanically except high-grade dark tea, which plucked by hand. In addition, the requirements of plucking criterion differ depending on the subtype of dark teas.

Table 6.2　　　　Dark tea grades and fresh leaves requirements

Dark Tea	Fresh Leaves
One-Level	One bud and three or four leaves
Two-Level	One bud and four or five leaves
Three-Level	One bud and five or six leaves
Four-Level	Banjhi plucking

After plucking, the fresh leaves should be transferred to the factory as soon as possible and be spread to the clean and hygienic floor (spreading thickness ≤ 30 cm) for partially moisture losing. This is beneficial for the subsequent fixing. Fresh tea leaves below the fourth grade or after the summer can be directly used for fixing without spreading.

Fixing—The purpose of dark tea fixing is the same as that of green tea, mainly for the inactivation of enzymes at high temperature. However, the fixing temperature of dark tea is higher than green tea. Furthermore, it is essential for dark tea to carry out "sprinkle water" before fixing since the fresh leaves are coarse, mature and lack of moisture. The "sprinkle water" should be uniform with the ratio of fresh leaves: water equal to 10 : 1 to avoid the uneven and poorly fixing.

The proper fixing standards: the young leaves are entangled, supple and sticky and the stems cannot be easily broken off; the color of the leaves turns from fresh green to dark green, and tea leaves show unique fragrance.

Rolling—The rolling of dark tea can be divided into two processes: initial rolling and re-rolling. When comes to initial rolling, it should be carried out quickly while hot, also called "hot rolling". The rolling process can shape the leaves into enveloped, leading to the breakage of cells, and then provides conditions for the piling. At the time of piling is completed, the tea leaves should be unraveled and re-rolling to make rolled tea sticks.

In general, medium-sized rolling machine is used for initial rolling while small-sized rolling machine is used for re-rolling process. In the case of avoiding "luffa ladle, peeled stem" during rolling, low pressure, short time and slow rolling speed should be applied.

Piling—As piling is a unique process in the production of primary dark tea, it plays a key role in the formation of special dark tea characteristics. The piling process has two purposes: ① To oxidize polyphenols and remove part of the astringent taste; ②To change leaves color from dark green or dark greenish yellow to yellowish brown. Under piling, the tea stack is covered with a damp cloth at a height of 70-100 cm at 25 ℃ to preserve the heat and moisture. The humidity is about 85%, and it is necessary to put the pile at a well-ventilated backlighting place for 12-24 h. The tea leaves are piled and lead to a series of oxidation, condensation and degradation of tea chemical compounds under the synergic effects of hyther and microorganisms. Under that condition, the polyphenol content is decreased by 20%-25%, thereby forming a unique flavor of dark tea. Additionally, it is necessary to ensure that the moisture content of the tea leaves is about 65% when piling. If the moisture is too high, the leaves are very easily to decay while if the moisture is too low, the piling process is slow and the conversion is not even.

Proper piling standards: As the tea leaves are piled around 24 h, the hand feels warm when reaches into the pile. The temperature of the leaves achieve about 45 ℃. If the surfaces of the pile appear droplets with yellowish brown leaves and give off vinasse smell or sour and acrid smell, the tea stack should be unraveled immediately and re-rolling. However, it is noteworthy that the different subtype of dark teas would use different piling strategy.

Insufficient piling results in the yellow and green leaves and thick astringent taste, whereas the excessive piling show leaves with dark color and slippery feeling. Factors affecting the piling: tea stack tightness, moisture content, leaf temperature, ambient temperature, oxygen level, etc.

Drying—The drying process of dark tea is the same as that of other teas, mainly to lose moisture, consolidate the quality, and further enhance its unique flavor. However, different dark tea producing regions have different drying methods. At present, there are several major drying methods:

i. Sun drying. Sunlight not only removes the moisture, but also promotes the physical and chemical changes of tea components. The appropriate moisture content is 13%.

ii. Open fire drying. Pine wood is commonly used for open flame drying. It can remove the moisture and produces a special smoke flavor under the smoked baking of pine wood. The appropriate moisture content is 8%-10%.

iii. Mechanical drying. At present, it is the most common method used to ensure the product quality, hygiene and safety requirement, and greatly improves production efficiency. The appropriate moisture content is 10%.

3. Key microorganisms in dark tea processing

The formation of unique quality of dark tea and special functional ingredients are mainly the systematic result of the enzymatic reaction and hyther with the presence of microorganisms. Microorganisms promote the quality components formation through the combined effects of biochemical motility (ectoenzyme), physicochemical power (microbial heat) and microbial metabolism. The difference of dominant microbial communities and reaction activities result in the different quality characteristics of various dark teas (Table 6.3).

Table 6.3　Dominant microbiota and main quality characteristics of different dark tea

Dark tea type	Dominant microbiota	Quality characteristics
Fuzhuan brick tea	*Eurotium cristatum* (golden flower)	Tea infusion orange/orange red, showing golden flowers, special "arohid flavour"
Puerh tea	*Aspergillus niger*, Yeast	Tea infusion reddish brown/dark red, superior stale and mellow
Qingzhuan brick tea	Filamentous fungi	Tea infusion orange red, appearance blue-auburn, superior pure aroma
Kangzhuan brick tea	*Aspergillus niger*, *Penicillium*, Yeast	Tea infusion red, appearance brown-auburn, and superior mild aroma
Liubao tea	Aspergillus, Penicillium, Rhizopus	Tea infusion red, superior dense aroma

词汇表

frontier nationalities [ˈfrʌntɪə(r) ˌnæʃəˈnælɪtiz] 边疆民族

banjhi plucking 开面采。芽梢停止生长形成驻芽的嫩梢称作"开面"

piling [ˈpaɪlɪŋ] v. 渥堆

sprinkle waters 洒水

supple [ˈsʌpl] adj. 柔韧性好的；易弯曲的；柔韧的

initial rolling 初揉

re-rolling 复揉

hot rolling 热揉。但当高温加热杀青后，叶片受湿热的综合作用，多糖分解，细胞膨压降低，原果胶物质在水热作用下，部分分解为水溶果胶，并带有一定程度的黏着性。采取趁热揉捻，能塑造出良好的外形

unravel [ʌnˈræv(ə)l] v. 解块

luffa ladle, peeled stem 丝瓜瓢，脱皮梗

> backlighting place 背光处
>
> hyther [ˈhaɪθə] *n.* 湿热作用
>
> vinasse, acrid smell 酒糟气，酸辣气。生产中当嗅到茶坯有酒糟香和轻微酸辣味时，渥堆已达适度
>
> dominant microbiota 优势微生物群
>
> ectoenzyme [ˌektəʊˈenzaɪm] *n.* 胞外酶。指细胞内合成而在细胞外起作用的酶，如多酚氧化酶、果胶酶、纤维素酶、蛋白酶等
>
> microbial heat 微生物热。微生物在其自身繁殖生长活动中，自身生长代谢必然释放大量的热能，这种热能既是自身活动的需要，又使渥堆堆温迅速升高，加速了湿热作用和酶促反应速率，为黑茶品质形成奠定了能量基础
>
> *Eurotium cristatum* 金花，一种黄色颗粒状菌体，学名冠突散囊菌，是茯砖茶发酵过程中的优势菌种，其从茶叶中获取营养物质，通过自身代谢，产生多种酶催化茶叶中各种物质发生转化，形成了茯砖茶特有的色、香、味品质，故其是评价茯砖茶品质的重要指标
>
> *Aspergillus niger* [ˌæspəˈdʒɪləs ˈnaɪdʒə] 黑曲霉
>
> Filamentous fungi [filəˈmentəs ˈfʌŋgaɪ] 丝状真菌
>
> Penicillium [ˌpenɪˈsɪlɪəm] 青霉
>
> Rhizopus [ˈraɪzəʊpəs] 根霉

思考题

1. 将下面的句子译成英文。

（1）黑茶加工工艺主要包括杀青、揉捻、渥堆、干燥，其中渥堆是黑茶加工中至关重要的工序，是形成其独特品质的关键。

（2）黑茶加工中，在微生物胞外酶的作用下产生独特的生化物质，在脂质代谢与肠胃调节等方面具有突出作用。

（3）黑茶加工过程中杀青适度的标准是嫩叶缠绕。叶软带黏性，茶梗不易折断，叶色由青绿转为暗绿，散发茶香。

2. 将下面的句子译成中文。

（1）Piling is a unique process in the primary production of dark tea, and also a key process in the formation of dark-wool tea quality.

（2）Leaves are yellow and green, thick and astringent taste, it means that the piling is insufficient; Leaves are dark and slippery when holding the green tea with a heavy sour smell, the piling is excessive.

（3）Microorganism promote the complex changes in the quality of dark tea through the combined effects of biochemical motility (ectoenzyme), physicochemical power (microbial heat) and microbial metabolism.

参考文献

[1] 张大春,王登良,郭勤. 黑茶渥堆作用研究进展[J]. 中国茶叶,2002(5):6-8.

[2] 胥伟,吴丹,姜依何,等. 黑茶微生物研究:从群落组成到安全分析[J]. 食品安全质量检测学报,2016(9):3541-3552.

[3] WANG C C, SHIANG J C, CHEN J T, et al. Syndrome of inappropriate secretion of antidiuretic hormone associated with localized herpes zoster ophthalmicus[J]. Journal of General Internal Medicine,2011,26(2):216-220.

[4] ZHENG W J, WAN X C, BAO G H. Brick dark tea: a review of the manufacture, chemical constituents and bioconversion of the major chemical components during fermentation [J]. Phytochemistry Reviews,2015,14(3):499-523.

[5] ZHANG L, DENG W, WAN X. Advantage of LC-MS metabolomics to identify marker compounds in two types of Chinese dark tea after different post-fermentation processes [J]. Food Science and Biotechnology,2014,23(2):355-360.

[6] ZOU Y, QI G, CHEN S, et al. A simple method based on image processing to estimate primary pigment levels of Sichuan Dark Tea during post-fermentation[J]. European Food Research and Technology,2014,239(2):357-363.

Lesson 7 Yellow Tea

Yellow tea is a lightly fermented tea with a mellow and brisk taste, which has a unique processing called "yellowing". The process of yellow tea is similar with green tea.

1. Introduction

The yellow tea has typical yellow color of tea leaves and infusions, attributed by the unique process of yellowing. Basically, yellow tea processing turns green tea leaves to yellow and promotes the formation of yellow tea quality.

In China, the annually production of yellow tea is approximately 8,000 tons, mainly for domestic consumption. It is a special tea with its history dates back to the 16th century. The main growing areas are Anhui, Hunan, Hubei and Sichuan provinces. The leaves are rolled to be curl during manufacturing. The yellow tea can be divided into at least three categories according to the plucking standards and tenderness of the leaves. Huangyacha, or Bud yellow tea, is processed from buds. Huangxiaocha, or small-leaf yellow tea, is prepared from one bud and one or two leaves. Huangdacha, or large-leaf yellow tea, is produced with one bud and three or four tea leaves.

2. Processing

Fixing—Fixing process is the foundation for the formation of yellow tea characteristics. At first, the high temperature is used to inactivate the enzymes. Based on this, other processes are applied to transform the components and form the unique color, aroma and taste of the yellow tea. However, if the fixed fresh leaves appear red, it is harmful to the yellow tea quality. In order to facilitate the formation of yellow tea quality compounds, the fixing temperature of the fresh leaves is relatively lower than that of green tea. Under the damp heat condition, the more degradation to chlorophyll and automatic oxidation/isomerization of polyphenolic compounds are observed. Additionally, the hydrolysis of starch into monosaccharides and proteins into amino acids, etc., also promote the formation of yellow tea quality characteristics.

The manufacturing of yellow tea is almost similar to that of green tea, from the plucking of the fresh leaves to the final drying. At the time of the fresh leaves are plucked and spread out for a short time, the manufacture process begins. The next process must be carried out on the same day as the leaves are plucked. The temperature of pan fixing is about 150℃, which is lower than that of green tea. This is favorable for yellowing and controls the moisture not evaporating quickly from leaves.

Yellowing—Yellowing is the key and characteristic process for the formation of yellow tea quality. The yellowing is carried out on the basis of inactivation of the enzyme activity and inhibition of the enzymatic oxidation of the polyphenol compound. Although effects are made from the beginning of the fixing to the end of the drying to promote the characteristics of yellow tea, the most significant process of the yellow tea characteristics forming is still yellowing.

During the yellowing process, the leaves are stacked, and the polyphenols are oxidized because of warm condition. While the leaves are placed in a container or covered with a wet cloth, the temperature increased to accelerate the yellowing change by thermal instead of enzymatic reactions. There are two kinds of thermal reaction, one is damp heat, and another is dry heat. The damp leaf yellowing takes place after fixing, whereas the drying leaf yellowing occurs after initial drying. The yellowing time varies from several hours to 5 - 10 days, depending on the method used. Dry leaf yellowing requires more time than damp leaf yellowing and, as expected, a deeper yellow color is obtained. Therefore, the yellow tea leaves and infusion is a typical characteristic of yellow tea.

In the processing of yellow tea, both thermal reactions contribute to the unique quality of yellow tea. On the one hand, the damp and heat environment causes a series of oxidation and hydrolysis changes in the leaves, which is the dominant factor of the formation of yellow leaf yellow infusion and the mellow taste. On the other hand, the dry heat acts to enhance the intensity of aroma for yellow tea.

Drying—Generally, the drying process of yellow tea is carried out in several times. There are oven drying and pan drying. The drying temperature of yellow tea lower than green tea with the trend of low first then increase to high temperature, this is benefit to the yellowing of the leaves. In fact, low-temperature drying at the beginning lower down the evaporation rate of moisture, creating the condition of damp and heat to make the tea leaves slowly dry and dehydrated. Furthermore, the components are slowly converted under damp and heat condition and enhance the yellowing process with a significant change in color, aroma and taste. After that, the high-temperature drying is used to fix the quality of the yellow tea.

词汇表

yellowing [ˈjeləʊɪŋ] v. 闷黄

yellow bud tea, Huangya cha 黄芽茶。按鲜叶原料的嫩度,黄茶又分为黄芽茶、黄小茶和黄大茶。君山银针、蒙顶黄芽属于黄芽茶

small-leaf yellow tea, Huangxiaocha 黄小茶。如霍山黄芽、沩山毛尖、北港毛尖、平阳黄汤、远安鹿苑茶等

large-leaf yellow tea, Huangdacha 黄大茶。如皖西黄大茶、广东大叶青茶

思考题

1. 将下面的句子译成英文。

(1) 黄茶的制作过程为鲜叶杀青、揉捻、闷黄、干燥。黄茶的杀青、揉捻、干燥等工序均与绿茶制法相似,其最重要的工序在于闷黄,该工艺利用高温杀青破坏酶的活性,其后多酚物质的氧化作用则是由于湿热作用引起,并产生一些有色物质。

(2) 形成黄茶品质的主导因素是热化作用。热化作用有两种:一种是在水分较多的情况下,以一定的温度作用之,称为湿热作用;另一种是在水分较少的情况下,以一定的温度作用之,称为干热作用。

(3) 黄茶的干燥分为烘干和炒干,并分次干燥。干燥温度在比绿茶干燥温度低,且先低后高,这是形成黄茶香味的重要因素。

2. 将下面的句子译成中文。

(1) Fixing is an important procedure in the yellow tea processing, which helps to eliminate the enzyme activity, promote chemical transformations of contained substances, evaporate water and reduce the grassy taste. It plays an important role in the late change of yellow tea and the quality of yellow tea products.

(2) During "sealed yellowing", chlorophyll and polyphenols contained in fresh leaves undergo oxidation, cracking and transformation under the effects of heat and humidity, resulting in the yellow appearance.

(3) Two different drying techniques are often employed: baked drying and fried drying. The temperature during the drying of yellow tea is usually lower than that of other teas. The

temperature is usually low at the beginning, and then increased later. By this means, the moisture dispersion is slow, and the yellow tea can continue the "sealed yellowing" process while drying under hot and humid conditions.

参考文献

[1] 张娇, 梁壮仙, 张拓, 等. 黄茶加工中主要品质成分的动态变化 [J/OL]. 食品科学, http://kns.cnki.net/kcms/detail/11.2206.ts.20181213.1441.056.html.

[2] 范方媛, 杨晓蕾, 龚淑英, 等. 闷黄工艺因子对黄茶品质及滋味化学组分的影响研究 [J]. 茶叶科学, 2019, 39(1):63-73.

[3] KUJAWSKA M, EWERTOWSKA M, IGNATOWICZ E, et al. Evaluation of safety and antioxidant activity of yellow tea (*Camellia sinensis*) extract for application in food [J]. Journal of Medicinal Food, 2016, 19(3):330-336.

[4] ZHOU J, ZHANG L, MENG Q, et al. Roasting process improves the hypoglycemic effect of large-leaf yellow tea infusion by enhancing levels of the epimerized catechins that inhibit α-glucosidase [J]. Food & Function, 2018, 9:5162.

[5] XU J, WANG M, ZHAO J, et al. Yellow tea (*Camellia sinensis* L.), a promising Chinese tea: Processing, chemical constituents and health benefits [J]. Food Research International, 2018, 107: 567-577.

[6] GUO X, HO C T, SCHWAB W, et al. Aroma compositions of Large-leaf yellow tea and potential effect of theanine on volatile formation in tea [J]. Food Chemistry, 2018, 280: 73-82.

Lesson 8　Further Processing

Further processing of tea refers to the application of modern physical, chemical and biological high-tech to produce and process tea or tea active ingredients using the fresh tea leaves, semi-finished tea, finished tea or tea by-products. Further processing and comprehensive utilization of tea resources are considered as efficient way to utilize the middle and low-grade tea resources as therefore adding value, improving the utilization efficiency and creating new industries in tea. The further processing industry in China has more than 20 years developing experience, which can be categorized as follows: Tea functional ingredients extract; Instant tea series (solid beverage); bottled or canned tea beverage (Ready-to-drink, RTD); Tea-containing food products; Tea-containing health products and medicines; Tea-containing personal care products and daily necessities; Animal feed and health products; Plant protection agent, etc. The further processing products from tea and their diversity all derived from the common material basis, which are instant tea powder or concentrated tea juice, tea functional ingredients, and ultra-fine tea powder. Therefore, the extraction technology of functional components in tea directly restricts the development of further processing industry.

1. Ready-to-drink (RTD) teas

RTD tea, also known as iced tea, is growing faster than any other beverage category. Many RTD products are made from tea extract powders, similar to instant teas. RTD tea are available in different packaging includes canned, glass bottle, PET bottle and aseptic cartons. The global RTD tea market in 2016 reached 37 million tons and was valued at approximately US$ 40 billion, and sales are predicted to increase by 4.5 percent annually between 2019 and 2021. China is the world's largest producer and consumer of RTD tea, with tea beverage consumption reaching 14.4 million tons in 2017.

Rising health awareness, introduction of functional ingredients, rising disposable income and fast-paced lifestyle are some of the major driving force for RTD tea. Tea offers a healthy solution and RTD Tea is a convenient format. Consumers are embracing RTD teas as tasty, refreshing grab-and-go alternatives to traditional sodas, juices and water. It is easy to see exactly what's inside the bottle, which meets consumers' demands for transparency. RTD tea has a huge variety of flavours including jasmine, ginseng, herbal, lemon, milk, and honey etc.

To accommodate the increased interest in the benefits of tea, beverage makers are producing a variety of tea drinks using different types of tea, such as black, green, oolong, white, dark and yellow. However, the RTD green tea and RTD black tea are the most popular in the world. There are recent trends indicating that a more authentic-tasting tea as well as the less sweet and functional teas, which include vitamins, additional epigallocatechin gallate (EGCG), grape seed extracts, antioxidants, non-tea flavors such as peach, raspberry or superfruits, are pushing the consumption rate up as well. Composition of RTD teas according to their nutrition fact labels is exemplified in Table 6.4.

Table 6.4　Composition of ready-to-drink flavored-colored commercial teas

Tea flavor	Composition reported in label
Blueberry	Water, white tea extract (*Camellia sinensis*), citric acid, sodium citrate, sucralose, colorant (anthocyanins and carmine), white tea flavor, blueberry flavor, sodium 100 mg/L, and polyphenols 125 mg/L.
Citrus	Water, green tea extract (*Camellia sinensis*), citric acid, sodium citrate, concentrated lemon juice, colorant (class IV caramel), green tea flavor, lemon flavor, sodium 150 mg/L, and polyphenols 225 mg/L.
Lemon	Water, black tea extract (*Camellia sinensis*), citric acid, sodium citrate, sucralose, concentrated lemon juice, colorant (class IV caramel and anthocyanins), black tea flavor, lemon flavor, sodium 150 mg/L, and polyphenols 125 mg/L.

RTD tea is a ready prepared tea, produce either from tea extracts in the form of reconstituented dried powder or as freshly brewed extract from tea or fresh tea leaves. The general processing of RTD tea from freshly brewed extract includes tea + water→extraction→ clarification→ stabilization → mixing with ingredients & sweeteners → post - mixing → TAB (thermoacidophilic bacteria) removal→pasteurization→bottling or canning.

During RTD processing, formation of haze and tea cream is a typical phenomenon that occurs in the final product. This phenomenon causes healthy properties reduction and a concern in its storage. Several processes are then proposed to reduce tea cream formation in RTD tea beverages and instant tea. Enzymatic treatment with tea materials have been attempted by tannase and cell wall digesting enzymes (cellulases, hemicellulase, xylanases, proteases, pectinase etc.) to improve its quality in terms of stability, clarity and cold water solubility as well as extraction yield. After extraction, the centrifugation and/or filtration after adjustments in temperature are usually used to clarify tea slurry. The membrane filtration technology as microfiltration (MF) and ultrafiltration (UF) have been applied to produce RTD tea with authentic tea flavor. The Spinning Cone Column (SCC), widely acknowledged as the world's premier flavor - recovery technology, is applied to capture and preserve the volatiles during extraction. The recovered volatiles could be added to the mixture at a later stage.

2. Instant tea

Instant tea is solid and water-soluble powder, which produced from traditional tea or tea by-product by extraction, clarification, concentration, conversion, and drying, etc. These are sold as powders which require water to become reconstituted into a tea beverage. In general, it has a nature of good time conservation and remains its original tea flavor after re-constituted in water. The development of instant tea in China began in the early 1960s. Increasing preference in tea beverages such as domestic tea beverages and milk tea and the rapidly expansion of membrane filtration and concentration technology have fueled the Chinese instant tea market. In 2018, the total production of instant tea in China was nearly 20,000 tons.

In the production of instant tea, the basic processing flowchart, including raw materials, quality and yield, is shown in Figure 6.6. According to the difference of processing materials, instant tea can be divided into instant green tea, instant black tea, instant oolong tea, instant dark tea (including Pu-erh tea), instant white tea, instant yellow tea, instant jasmine tea, etc. In China, the development of instant tea processing technology has experienced the technical systems of tank extraction, plate and frame filtration (or centrifugal filtration), melt dissolution, vacuum concentration, and vacuum drying or spray drying before the 1990s. The mainstream of advanced processing technology systems has boosted in the 21st century, comprising advanced countercurrent extraction, ultrafiltration (microfiltration), membrane filtration, reverse osmosis membrane concentration, spray drying or freeze drying. With the rising health awareness and the technological advancements, the integration of advanced technology to produce instant tea with the characteristics of good solubility (cold or hot melt), clear and transparent, high aroma, no bitter taste, high-quality tea infusion color, has become the main development trend of instant tea production and consumption.

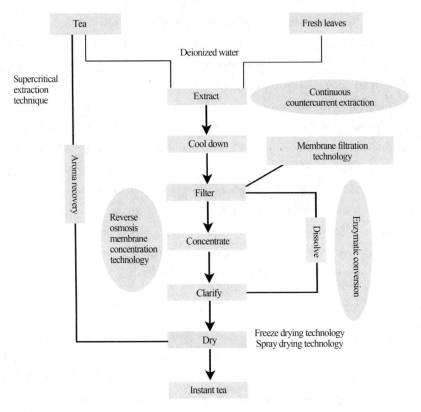

Figure 6.6　Instant tea processing

3. Development and utilization of tea functional components

Researches into the functional ingredients of tea started about 30 years ago in China. Technological innovation on the extraction, separation and purification of active components from tea, especially polyphenols, has promoted the global development of tea comprehensive processing (the extraction of active ingredients, such as instant tea, tea polyphenols, caffeine, theanine and theaflavins from tea leaves to get the soluble extracts of tea leaves) and tea extracts applied in the field of health. Over the past 20 years, Chinese tea scientists and technicians have been devoting themselves to the study of new technique for extraction, separation and purification of tea functional components such as catechin, theanine and theaflavin, which has effectively improved the technology level of tea further processing and expanded the global consumer market.

In the traditional extraction process of tea polyphenols, organic solvents such as ethyl acetate, chloroform or dichloromethane are used as a separation and purification medium, and the product has a safety risk due to solvent residue. The current state-of-the-art process applies macroporous adsorption resin column chromatography technology to the industrial separation of tea catechins, and utilizes the selectivity of macroporous resins to achieve separation and purification of mixtures. The use of water and edible alcohol as a solvent to efficiently separate and purify the catechin components and safely remove caffeine in green, solves the problem of solvent residues such as ethyl acetate, methylene chloride and chloroform in the conventional process, and the safety of catechin products was greatly improved. At the same time, through the application of membrane technology, especially the molecular sieve effect of nano-membrane, we can effectively control the proportion of various functional components in tea extract, and develop various special specifications of tea extract products.

1) Innovative separation and purification technology of tea catechins monomers

By using enzyme engineering technology and column chromatography on-line monitoring technology, the bottleneck of efficient separation and preparation of catechin monomers has been broken through, and the preparation of EGCG or EC monomer has been industrialized and scaled at the gram level of laboratory. This technology supports the annual production capacity of 90%-98% of EGCG or EC monomers in China to reach over 30 tons, which leads to the emergence of catechin monomers as Active Pharmaceutical Ingredients (API) natural drugs, dietary supplements, health foods and natural cosmetics in the world.

2) Industrial production of theaflavins by enzymatic oxidation of tea catechins

Because of the low content of theaflavins in black tea, the cost of extracting theaflavins from black tea is high, so it is difficult to realize industrialization to meet the international market demand for health products. Researchers screened polyphenol oxidase which can efficiently express and catalyze catechin oxidation from plant (tea, pear, apple...) and microbial sources, and constructed a new technology for the preparation of theaflavins by enzymatic oxidation of green tea catechin. Through optimizing the combination of catechin substrates of different high esterification ratios and polyphenol oxidase isoenzymes from different sources, the directional regulation of theaflavins component proportion was realized, and the large-scale separation and purification technology system of theaflavins monomer and high purity theaflavins was established. This technology has provided a feasible opportunity for the development of healthy products with theaflavins as active ingredients in the world.

3) Separation and purification of natural L-theanine

The antidepressant, sedative and sleep-enhancing effects of L-theanine attracted worldwide attention in the late 1990s. At present, the separation and purification of natural L-theanine from the aqueous layer after extraction of tea polyphenols has been successful and industrialized. Large-scale development of L-theanine with a purity of 20% to 98% meets the needs of healthy food development in the international market.

词汇表

ready-to-drink（RTD）teas 即饮茶

instant tea 速溶茶

ultra-fine tea powder 超微茶粉

polyethylene terephthalate（PET）聚对苯二甲酸乙二醇酯

health awareness 健康意识

disposable income 可支配收入

fast-paced lifestyle 快节奏的生活方式

grab-and-go 拿上就走；即拿即走

authentic-tasting tea 原味茶

thermoacidophilic bacteria（TAB）嗜热嗜酸菌

Pasteurization [ˌpæstəraɪˈzeɪʃən] n. 巴氏灭菌(消毒)法

tannase [ˈtæneɪs] 单宁酶

cellulose [ˈseljʊleɪs] 纤维素酶

hemicellulase [hemɪˈseljʊleɪs] 半纤维素酶

xylanase [ˈzaɪləneɪs] 木聚糖酶

protease [ˈprəʊtieɪz] 蛋白酶

pectinase [pekˈtiːnəz] 果胶酶

microfiltration（MF）and ultrafiltration（UF）微滤和超滤

Spinning Cone Column（SCC）旋转锥体柱

state-of-the-art adj. 使用最先进技术的；体现最高水平的

Active Pharmaceutical Ingredients 原料药；活性药物成分

dietary supplements [ˈdaɪətəri ˈsʌplɪments] 膳食补充剂；保健食品；营养保健品

realize industrialization 实现工业化生产

cream processing 转溶，速溶茶生产工艺。转溶就是根据茶乳酪物质的组成、理化性质，采用添加酶、酸或碱等方法，将这些不溶物转化为可溶物的过程，使速溶茶具有较好的溶解性，尤其是冷溶性

macroporous adsorbent resin 大孔吸附树脂

membrane separation 膜分离，膜是具有选择性分离功能的材料。利用膜的选择性分离实现料液的不同组分的分离、纯化、浓缩的过程，这过程是一种物理过程，不需发生相的变化和添加助剂

思考题

1. 将下面的句子译成英文。

（1）茶叶深加工的产品开发已突破农产品范畴，向食品、饮料、医药、轻工、建材等领域广泛延伸与拓展。

（2）速溶茶作为被称为国饮的传统茶叶的深加工形式，以其方便、营养、可口、健康

的特点为越来越多的消费者接受并喜爱。

（3）借助深加工技术改变传统茶叶的感官品质和产品形式，可以满足不同消费者需求，是扩大消费群体和扩展产品市场的重要手段之一。

2. 将下面的句子译成中文。

（1）The use of high-efficiency, low-consumption, low-emission green technology and comprehensive extraction of active substances in tea is essential for solving the problems associated with the costs and benefits of tea extract.

（2）Extraction is a basic and important technological step in tea deep processing, its efficiency directly affects the yield and product quality of tea components which are easy to polymerize under relatively high temperature and the formation of the tea complex do not dissolve well in cold water.

（3）With further studies of the relationships between tea active ingredients and human health, the applications of tea functional ingredients have been greatly expanded; the quality specifications of tea extracts are becoming more detailed and specific.

参考文献

[1] FLORES-MARTÍNEZ D, URÍAS-ORONA V, HERNÁNDEZ-GARCÍA L, et al. Physicochemical parameters, mineral composition, and nutraceutical properties of ready-to-drink flavored-colored commercial teas [J]. Journal of Chemistry, 2018, 286 (1541). https://doi.org/10.1155/2018/2861541.

[2] SUBRAMANIAN R, KUMAR C S, SHARMA P. Membrane clarification of tea extracts [J]. Critical Reviews in Food Science and Nutrition, 2014, 54(9): 1151-1157.

[3] KUMAR C S, SUBRAMANIAN R, RAO L J. Application of enzymes in the production of RTD black tea beverages: A review [J]. Critical Reviews in Food Science and Nutrition, 2013, 53(2):180-197.

[4] LIU Z H. The development process and trend of Chinese tea comprehensive processing industry [J]. Journal of Tea Science, 2019, 39(2): 115-122.

[5] LIU Z, GAO L, CHEN Z, et al. Leading progress on genomics, health benefits and utilization of tea resources in China [EB/OL]. Nature outlook: tea, 2019, 566 (7742). https://www.nature.com/articles/d42473-019-00032-8.

[6] VUONG Q V, GOLDING J B, NGUYEN M, et al. Extraction and isolation of catechins from tea [J]. Journal of Separation Science, 2015, 33(21):3415-3428.

[7] ALASALVAR C, PELVAN E, ÖZDEMIR K S, et al. Compositional, nutritional, and functional characteristics of instant teas produced from low- and high-quality black teas [J]. Journal of Agricultural and Food Chemistry, 2013, 61(31):7529-7536.

Lesson 9 Tea Storage

The processing from primary to refining and the trade from transshipment to consumer drinking are major processes of tea storage and transportation. The quality of tea should remain as fresh and flavorful as possible during storage. Enhanced tea stability during storage not only protect the rights and interests of consumers, but also give full play to the economic value of tea and improve the economic benefits of tea industry. This section mainly introduces the effects of external storage environment conditions on tea quality and the transformation of quality components during storage. In addition, the storage methods also mentioned.

1. Effects of storage environmental conditions on tea quality

Temperature—Temperature plays a more important role on tea quality (taste, color, aroma, etc.) change than oxygen and moisture content. As the ambient temperature increases, the content of active ingredients in tea gradually decreases. Changes in the content of aromatic substances, vitamin C, amino acids and tea polyphenols are all related to temperature. Therefore, cold storage (below than 10 ℃) is the most effective method for tea storage.

Humidity—Humidity refers to the relative humidity of the air in the storage environment, and also the moisture content of tea itself. During storage and transportation, the oxidation of the contained components, the occurrence of mold and the changes after mold are closely associated with humidity. Dry tea is extremely hygroscopic. If the relative humidity of the storage environment is too high, the tea will quickly absorb the moisture. Generally, the drier tea absorbs moisture more quickly. With the increase of moisture content in dry tea, various chemical reactions are gradually accelerated and the condition more suitable for mold breeding. Therefore, the higher the moisture content, the more pronounced the diffusion and interaction of the substances, the more the mold propagates, the faster the tea deteriorates. In general, the moisture content of tea leaves should not exceed 6%, the ambient relative humidity should below than 50%.

Illumination—Illumination can accelerate the oxidation of pigments and lipids in tea, especially for chlorophyll, which is easily degraded under light irradiation. Therefore, light has a negative impact on the preservation of tea, leading to the degradation of polyphenols, amino acids and alkaloids and acceleration the aging of tea. The reduction of polyphenols, amino acids, and alkaloids are the factor resulting in the decline of tea quality including color, infusion and aroma, etc.

Oxygen—During the storage of dry tea, the ingredients of tea are auto-oxidized with the presence of oxygen. The aldehydes, hydrazines, phenols, etc., are easily oxidized by the molecular oxygen in the air to form oxides. As such oxides are present alone or in combination with other substances, they can continue to oxidize and decompose, this is the basis of food oxidation. Tea polyphenols, vitamin C, aroma components, etc. in dry tea can be automatically oxidized. Most of the tea products after oxidation are unfavorable for quality except the post-ripening dark tea.

2. Changes in the composition of tea during storage

Polyphenols—The change of tea polyphenols during storage is mainly non–enzymatic auto–oxidation, which means the oxidation of tea polyphenols occurred under certain temperature and humidity. Although this oxidation is not as intense as enzymatic oxidation, long-term storage can lead to significant deterioration of tea quality. Especially for teas with high moisture content at relatively high temperature, the reduction of tea polyphenol content during storage is more significant.

Chlorophyll—Chlorophyll, accounting for 0.7%–1.2% of dry tea, is the main pigment that constitutes the appearance of green tea and the infused leaves. Chlorophyll is extremely unstable under light and heat due to the reaction of replacement and decomposition, which makes the emerald green chlorophyll demagnesize and turn into brown pheophytin. The magnesium removal reaction become stronger as the moisture content increased. When more than 70% the chlorophyll in green tea is converted to pheophytin, the green tea is obviously browned with the yellow appearance and yellowish brown infusion.

Amino acids—The content of amino acids is gradually reduced during the storage. On the one hand, amino acids can react with polyphenols and soluble sugars to form polymers. On the other hand, amino acids can be oxidized, degraded and converted under certain temperature. As the storage time extends, the amino acid degrades and further affects the quality of the tea.

Vitamin C—During the storage, the reduced vitamin C in tea is converted into oxidized vitamin C due to the oxidation, which not only causes the decrease of tea nutritional value, but also leads to the brown appearance and brown tea infusion. As the content of vitamin C drops by more than 15%, the tea quality deteriorates obviously.

Aromatic volatiles—During the storage process, the content of aroma components in the tea leaves is significantly reduced with the loss of fresh taste and the presence of stale smell, and the tea leaves even produce an unpleasant smell for improper storage.

3. Tea storage

A thorough understanding of the quality deterioration factors during tea storage will be beneficial for finding the correct storage and transportation conditions to extend shelf life of tea. Although the storage methods of different teas are slightly different, they are basically stored in an environment away from light, humidity and high temperature. Low temperature storage is best for all teas except for white tea and dark tea which can also be stored at room temperature.

Store tea in a dry environment—Keep teas in a dry container and stay away from moisture. Dampness and humidity can dramatically reduce the tea's lifespan, and could even cause mold.

Keep tea away from sunlight—Always keep tea sealed in an airtight container that blocks out all light. Glass jars are not acceptable for tea storage because the sunlight oxidize the substances of tea over time. Keep the tea in a dark and cool place, like a shelf or drawer out of the sunlight because sunlight will produce heat then weaken tea fragrance and change tea quality.

Keep tea sealed from the air—If tea is sealed in a pouch, make sure that there is as little air as possible when you close the pouch. If you have a vacuum sealer, vacuum-sealed foil bags have the longest shelf life of tea but are not suitable for every tea. Long leaves that are not curled and rolled can be easily crushed under a vacuum. In these cases, make sure to keep your bags sealed between uses and remove excess air.

Store green tea in the refrigerator—If you have a vacuum sealer, you can feel free to thoroughly seal your tea in a moisture-free environment, and store it in the refrigerator on a colder setting, or the freezer. This is controversy because improperly-sealed tea stored in a freezer will be damaged. Besides, it is better to store your tea with proper amount as the package should not be opened too often.

4. Summary

The deterioration of tea quality is generally due to various content changes of the components during storage, such as tea polyphenols, amino acids, vitamin C, lipids and pigment, which lead to the loss of aroma and original unique taste. The factors affecting the deterioration of tea are mainly tea moisture content, storage temperature, oxygen level and light. In addition, packaging materials and microorganisms can also affect the tea quality. Therefore, it should be considered comprehensively for all the factors in the preservation and storage of tea. For green tea, the storage conditions of low temperature, drying, dark and oxygen insulation can delay the deterioration of green tea quality to the greatest extent, while new packaging, storage and processing technology can also be used to prolong the storage stability of tea.

词汇表

moisture content 含水量
ambient temperature 环境温度
cold storage 低温贮藏
hygroscopic [haɪgrəˈskɔpɪk] *adj.* 吸湿的
pheophytin [fiːəˈfaɪtɪn] *n.* 脱镁叶绿素
degradation [ˌdegrəˈdeɪʃn] *n.* 降解
deterioration [dɪˌtɪərɪəˈreɪʃn] *n.* 劣变，变坏
stale smell 陈味

思考题

1. 将下面的句子译成英文。

(1) 干茶色泽的变化,与温度、含水量的关系较大,茶叶干燥和冷藏是防止茶叶变质的良好方法。

(2) 随着人们生活水平的不断提高,消费者对茶叶包装的质量要求越加严格,传统的简易包装、通用包装或散装零售等包装形式已不能适应当前茶业发展的技术需求。

(3) 茶叶保鲜是一项综合技术措施,任何单一保鲜技术都有其优缺点,只有将各项技术综合运用,相互补充才能取得更好的效果。

2. 将下面的句子译成中文。

(1) Different teas have different shelf lives and you need to protect them from common elements that could easily ruin a perfectly good tea.

(2) When the storage time is longer, the aroma becomes fainter and the tea quality is inferior, so tea quality can be evaluated by detecting the changes of its aroma during storage.

(3) Extended storage, being necessary to develop desirable taste and aroma of ripen pu-erh and aged oolong tea, can lead to loss of quality in white, yellow, green and black tea.

(4) The storage time could affect the quality of the tea, which could decrease and many of the beneficial components could be gradually lost during storage. When the quality of the tea decreases and beneficial components are lost, the aroma of the green tea will change.

参考文献

[1] KOSIŃSKA A, ANDLAUER W. Antioxidant capacity of tea: Effect of processing and storage[M]//PREEDY V.Processing & Impact on Antioxidants in Beverages. Oxford, U.K.: Academic Press, 2014: 109-120.

[2] FRIEDMAN M, LEVIN C E, LEE S U, et al. Stability of green tea catechins in commercial tea leaves during storage for 6 months [J]. Journal of Food Science, 2009, 74 (2):5.

[3] STAGG G V. Chemical changes occurring during the storage of black tea [J]. Journal of the Science of Food & Agriculture, 2010, 25(8):1015-1034.

UNIT Seven Tea Quality Evaluation

Lesson 1 Water Quality

Tea is a worldwide beverage. Water quality affects tea taste aroma as well as health conditions. It is pointed out that, using purified water rather than tap water or natural water, could improve the quality of tea infusion owing to different elements and mineral matter contents. There may be certain relationships between water quality and the extractions of tea polyphenols, amino acids, saccharides, caffeine and other components, which influence the quality of tea infusion. In general, most tea brews best in moderate to soft water.

1. Water and tea

China has a tradition of selecting the optimum water type for individual teas. The subtleties of well water, spring water and river water were well known to early tea masters. Lu Yu, China's legendary tea connoisseur, maintained that the live water of mountain springs imparted a superior and subtler flavour to tea. Su Shi, a famous scholar of the Song dynasty (960-1279 AD), maintained that live water should be boiled by a live fire. A Ming Dynasty document, Written Conversation of Plum Blossom Herbal Hall, crystallizes the essential importance of water for tea. It states: "The inherent quality of tea must be expressed in water. When a tea that is an eight points meets with water that is a ten points, the tea is also a ten points! When water that is an eight points pairs with a tea that is a ten points then the tea is just an eight points." This premise is very true: water of the highest quality expresses the best qualities of a tea infusion. Therefore, when superior quality tea is brewed using inferior water, the resulting infusion is inferior. It is interestingly reported recently that locally grown tea matched to local water tastes best; since the tea trees are watered and nourished by the same water and soil. This is the same idea as in the common saying: Dragon Well Tea must be brewed with the water of Tiger Run Spring. This same principle, by extension, could also be applied to any tea.

Choice of water for tea use, based on taste and affinity for tea, is crucial; water is the substance of tea and helps to best express the tea's essence. For this reason, the ancients, in exacting and precise measure, evaluated and ranked water sources most suitable for tea use. Fresh, flowing water is also essential. If the water is flowing or moving then it stays oxygenated. Flowing water will then be fresh, not stagnant, providing a superior taste and improving tea infusion flavor. Any water, no matter the source, must be moving to stay oxygenated—flowing water is superior to stagnant water. Water that does not flow or loses oxygen tastes flat, imparting a flat taste to steeped tea.

Ancient Chinese classified water based on the apparent source: either from the sky or earth. Waters from the sky were called "heaven's spring". Sky waters included rain, snow, hail, frost, and dew. Earth waters included spring, river, and well water. However, most of these natural water sources are unsuitable for tea because of elevated hygiene concerns at present. As a rule of thumb, the usual tap water shall be filtered for tea to remove chlorine for improving the taste, because chlorine in the water can give a bitter taste.

In addition to the way tea tastes, water type also influences liquor colour. As a rule soft water imparts an attractive, bright, clear reddish hue, however discoloration of tea liquors occurs when using predominantly temporary hard water or chalybeate waters–the latter being impregnated with iron salts. Hard water tends to give a duller, darker colour to the liquor. This colour variation is attributed to alkaline pH, higher calcium and other minerals, etc. One recent publication pointed out that calcium ion has great influence on the sensory quality of green tea infusion, where it was found to reduce the umami sweet, bitter tastes and increase the astringent taste. Moreover, magnesium and calcium in hard water promoted two undesirable outcomes of the tea brewing: tea cream and scum formation. Tea cream is the precipitate matter that forms as tea cools and is caused reaction between caffeine and tea flavanols. Tea scum is a surface film that forms on the tea infusion surface, composed of calcium, hydrogen carbonates and other organic material. Tea scum happens particularly when tea is brewed with hard water. The film occurs due to calcium carbonate triggering oxidation of organic compounds.

2. Water boiling visual clues

Lu Yu delightfully visualized the four primary temperatures of water used for brewing tea as follows:

Column of steam steadily rising—This is the period during which a visible pillar of steam materializes, approximately 170 to 180 °F (72 to 82 ℃).

Fish eyes—This is when large lazy bubbles start to break the surface, approximately 180 to 200 °F (82 to 93 ℃).

String of pearls—This is the moment almost at the boil, when tiny bubbles appear to loop near the perimeter, approximately 190 to 200 °F (88 to 93 ℃).

Turbulent waters—This is a full rolling boil, when the water becomes highly oxygenated, approximately 200 to 212 °F (93 to 100 ℃).

词汇表

water quality 水质

tap water 自来水

mineral matter 矿物质

spring water 泉水

well water 井水

connoisseur [ˌkɒnəˈsɜː(r)] n. 鉴赏家；行家

premise [ˈpremɪs] n. 前提；假定

Live water should be boiled by a live fire 活水还需活火烹。选自北宋诗人苏轼写的一首七言律诗《汲江煎茶》，诗中描写了从取水、煎茶到饮茶的全过程

Written Conversation of Plum Blossom Herbal Hall《梅花草堂笔谈》。明代张大复撰写的随笔。其中谈及茶和水之妙，有一段非常精彩的论述。他说："茶性必发于水，八分之茶，遇十分之水，茶亦十分矣；八分之水，试十分之茶，茶只八分耳。"

Heaven's spring 天水。中国古人把天空的水包括雨，雪，冰雹，霜冻和露水统称为天水

Stagnant [ˈstægnənt] adj. (水或空气)不流动而污浊的；停滞的

flat taste [flæt teɪst] 味清淡

as a rule of thumb [æz ə ruːl ɒv θʌm] 根据经验

impregnate [ˈɪmpregneɪt] v. 使充满；使遍布

chlorine 氯

chalybeate water 铁质水

tea cream 茶乳酪

tea scum 茶浮沫

mineral water 矿泉水

distilled water 蒸馏水

deionized water 去离子水

思考题

1. 将下面的句子译成英文。

（1）中国传奇茶叶鉴赏大师陆羽认为，山泉的活水给茶叶带来了优越而微妙的味道。宋代（960—1279）著名学者苏轼认为活水还需活火烹。

（2）新鲜流动的水富氧，泡茶后茶汤味道优良，可有效提升茶汤风味。

（3）自来水经过滤去除氯气后泡茶，可有效改善茶汤口感。

2. 将下面的句子译成中文。

（1）A Ming Dynasty document, Written Conversation of Plum Blossom Herbal Hall, crystallizes the essential importance of water for tea. It states: "The inherent quality of tea must be expressed in water. When a tea that is an eight meets with water that is a ten, the tea is also a ten! When water that is an eight pairs with a tea that is a ten then the tea is just an eight."

(2) Choice of water for tea use, based on taste and affinity for tea, is crucial; water is the substance of tea and helps to best express the tea's essence. For this reason, the ancients, in exact and precise measure, evaluated and ranked water sources most suitable for tea use.

(3) Tea scum is a surface film that forms on the tea infusion surface, composed of calcium, hydrogen carbonates and other organic material. Tea scum happens particularly when tea is brewed with hard water.

参考文献

［1］ZHOU D, CHEN Y, NI D. Effect of water quality on the nutritional components and antioxidant activity of green tea extracts ［J］. Food Chemistry, 2009, 113(1):110-114.

［2］LI J, JOUNG H J, LEE I W, et al. The influence of different water types and brewing durations on the colloidal properties of green tea infusion ［J］. International Journal of Food Science & Technology, 2015, 50(11):2483-2489.

［3］ZHANG H, JIANG Y, LV Y, et al. Effect of water quality on the main components in Fuding white tea infusions ［J］. Journal of Food Science and Technology, 2017, 54(5):1206-1211.

［4］FRANKS M. The effect of water quality on green and black Tea ［D］. Ithaca, U.S.: Cornell University, 2018.

［5］PELTIER W. The ancient art of tea ［M］. Vermont, U.S.: Tuttle Publishing, 2011.

［6］ISO 3103-1980. Preparation of a liquor of tea for use in sensory tests ［S］.

［7］HEISS M L, HEISS R J. The story of tea a cultural history and drinking guide ［M］. New York, U.S.: Ten Speed Press, 476-480.

Lesson 2　Tea Utensils

As tea have traditionally been assessed organoleptically, i.e. with special reference to smell and taste, tea tasting has come to be considered more of an art than a science.

1. Professional tea tasting utensils

The basic utensils usually used for professional tea tasting, includes:

A tea tasting set—This includes a tasting bowl (white china tasting bowls) and a brewing cup (white porcelain cups) with its edge partly serrated or a half hole (Figure 7.1) and provided with a lid.

(1)　　　　　　　　　　(3)

Figure 7.1　A tasting set

扫码看彩图

A weighing scale—To measure accurately a similar portion of tea for each cup. This is usually 3 grams of tea. (The actual weighing out of tea samples on scales is bound more by ritual and tradition than scientific exactitude. Coins have long acted as the standard weight and average out at around 3 g or 2% of the brew. Tasters in Sri Lanka measure out the tea on the scales counterpoised by two 25 cent coins.)

A tasting spoon—Similar to a soup spoon.

A spittoon—To spit out the tea after tasting it.

A timer (or hourglass)—So that the tea may be steeped for a precise time of 3 – 5 minutes.

For the "lay taster" a couple of small teapots, cups, spoons and any watch or clock serves the purpose just as well.

According to ISO 3103 "Tea—preparation of a liquor for use in sensory tests", the basic equipment which a tea taster uses, includes:

Pot, of white porcelain or glazed earthenware, with its edge partly serrated (Figure 7.2) and provided with a lid, the skirt of which fits loosely inside the pot. The maximum capacities, when the pots are filled to the partly serrated edge, are (310 ± 8) mL for the large pot and (150 ± 4) mL for the small pot. The inside of the pot may be marked with a groove, or a coloured line, to indicate the volume of water which should be added. The lid should be loose-fitting and provided with a small hole to allow air to enter when the liquor is being poured out of the pot.

Bowl, of white porcelain or glazed earthenware. The maximum capacities are 380 mL for the large bowl and 200 mL for the small bowl.

NOTE: The diameter of the bowl is such that the pot and lid can rest inside to drain off the liquor, and the angle of the inside surface of the bowl is such as to allow the taster an uninterrupted view of the liquor without shadow. Various sizes of pot and bowl can be used, but it is recommended that one of the two sizes depicted in the Figure 7.2, should be adopted.

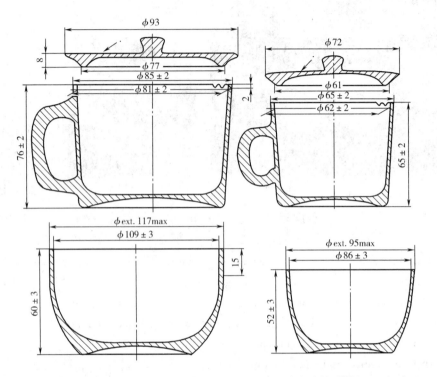

Figure 7.2 The size of tea pot and bowl recommended by ISO standard.

2. Daily tea utensils

Some of utensils for tea brewing is generally described in "the Chinese art of tea". Here, we specified two utensils used daily.

Tea pots, allowed the tea leaves in the teapot to be reinfused several times by successively adding more water, a method of tea brewing still followed in China. The first teapot began to appear toward the end of the Song dynasty, at a time when tea drinkers began to experiment with tea brewed from tea leaf. These early teapots were made from unglazed *zisha*, or purple sand clay (Figure 7.3). Teapots were given a lid, a handle and spout. Besides zisha teapot, there exist teapots made from porcelain, glass, silver and iron.

Gaiwan (**thin-walled, lidded cups with saucers**, Figure 7.4), comprises of three sections each with a meaning in Chinese culture: Saucer (Earth), Cup (People), and Lid (Heaven). The harmonious balance of the three sections helps with heat retention, avoiding spill, and ease of brewing. It is usually for brewing green teas. The most versatile Gaiwan is made from porcelain or glass. But they can be made from jade, or even Yixing clay. To use it, place some tea leaves in the cup and cover them with hot water. Then replace the lid and allow

Figure 7.3　Yixing tea pots

Figure 7.4　Gaiwan.

the tea to steep. Use the lid to stir the leaves and hold them back while you sip the tea. Always keep the saucer with the cup, using one hand to manage the lid and cup and the other to hold the saucer. When the tea is ready the Gaiwan should be covered and picked up on its plate with the left hand and placed on the up-turned fingers of the right hand. The lid should be positioned slightly askew and held in place with the thumb-just enough to allow the tea to pour out while retaining the leaves. Using two hands may at first appear to be easier, but it actually makes pouring more difficult, while using one hand locks the three pieces in place and holds them together as the gaiwan is inverted for pouring.

词汇表

assessed organoleptically/sensory evaluation 感官审评
a tea tasting set 茶叶审评工具组合
tea pot/brewing cup 审评杯
tea bowl/tasting bowl 审评碗
serrated [səˈreɪtɪd] 锯齿形
weighing scale 秤
tasting spoon 汤匙。茶叶审评中用于搅拌茶汤，舀取茶汤之用
spittoon [spɪˈtuːn] 水盂，茶叶审评中用来盛放废水和吐出的茶汤
timer/ hourglass 计时器，茶叶审评中用来冲泡计时
lay adj. 外行的；非专业的

思考题

1. 将下面的句子译成英文。

（1）将干茶样品放在审评盘里，检查茶叶等级、粒度、颜色、锋苗、紧细度、茶香和匀整度。

（2）审评杯盖上有一个小孔，以便在茶汤从杯中倒出时允许空气进入。

（3）审评碗内表面的设计角度可使审评人员能够观察茶汤的色泽，而不受阴影的影响。

2. 将下面的句子译成中文。

（1）The tea sensory evaluations demanded individuals who had their senses of sight, taste and smell developed to the highest possible level, since the job required not only a sharp eye, but also an equally delicate and discriminating nose and palate.

（2）The dry leaf of good tea will be free from stalks and twigs, and uniform in size. The former contribute worthless weight and sometimes off-flavors to the tea. If the tea is not uniform in size, it will not behave consistently from one brewing to the next.

（3）Pot, of white porcelain or glazed earthenware, with its edge partly serrated and provided with a lid, the skirt of which fits loosely inside the pot. The maximum capacities, when the pots are filled to the partly serrated edge, are (310 ± 8) mL for the large pot and (150 ± 4) mL for the small pot.

参考文献

[1] ISO 3103—1980. Preparation of a liquor of tea for use in sensory tests [S].

[2] HEISS M L, HEISS R J. The story of tea a cultural history and drinking guide [M]. New York, U.S.: Ten Speed Press, 2007.

Lesson 3 Formation of Tea Quality

Tea quality generally refers to the shape (appearance), the color, the aroma and the taste of tea. As a beverage, the aroma and taste of tea should be the core of quality. Meanwhile, with its commodity value, the appearance and glossy color could not be ignored neither. For sensory evaluation of tea quality, it contains the quality of tea shape (including the shape of dry tea and remained tea leaves after infusion), the quality of tea color (including the color of dry tea, tea liquor and infused tea leaves), the quality of tea aroma and tea taste. Accordingly, their formation principles are discussed in this chapter.

1. Tea shape formation

For varieties of tea, their shapes are of various postures, and most of them are certain artistic for appreciation. Even for remained tea leaves after infusion, there're postures like flower shape or complete leaves and so on. The tea shape is clear to touch and distinguish, it's an important item for tea quality determination.

Chemical composition affecting tea shape—The shape of tea is mainly determined by the tea making process, but also closely related to some of the included chemical constituents. The main ingredients related to the shape of tea are cellulose, hemicellulose, lignin, pectin, soluble sugar, moisture contents, total amount of soluble ingredients, etc. These components are related to the degree of tenderness of fresh tea materials, which would affect the flexibility and plasticity of tea texture and performance of tea making technology, thus further affect the quality of tea shape.

Main factors affecting tea shape—The diversification of tea shape is mainly related to ① the tea varieties, ② cultivation condition, ③ picking standards, ④ processing technology, and ⑤ storage process. They are not the decisive factors in the formation of tea shape, but considerable for the beauty of shape and the formation of quality.

Types of tea shape—Tea shapes could be divided into the following types according to different tea varieties and processing techniques: strip shape, curly shape, spherical shape, flat shape, needle shape, pointed shape, flower shape, granular shape, flake shape, powder form, sparrow tongue shape, annular hook shape, cluster shape, spiral shape, etc.

After infused, the tea leaves would absorb water to expand to size of the original fresh leaves. Then the shape of the remained tea leaves could be identified to the authenticity of tea, and related with the tea varieties, the cultivation conditions, and the observation of certain problems in the processing. These factors are important in the comprehensive analysis of tea characteristics.

The shapes of remained tea leaves after infusion could be roughly divided into: bud-shape, sparrow tongue shape, flower shape, whole leaf shape, half leaf shape, fragmentary shape, etc.

2. Tea liquor color formation

Chemical composition of tea color—Tea color includes the color of dry tea leaves, tea liquor and remained tea leaves after infusion. They are the comprehensive reflection of a series of degradation and oxidative polymerization of the chemical composition of tea in tea making process. The colored substances are mainly the flavonoids, flavonols (including anthocyanin) and its glycosides, carotenoids (including xanthine), chlorophyll and its transformation products, theaflavins, thearubigins, theabrownins, etc.

Types of tea liquor color—After a certain degree of infusion, the color characteristics of tea liquor could be divided into different types as follows: ①light green type, ②apricot green type, ③yellow-green type, ④apricot yellow type, ⑤ light yellow type, ⑥golden yellow type, ⑦orange-yellow type, ⑧orange red type, ⑨red bright type, ⑩red brilliant type, ⑪deep red type.

3. Tea aroma formation

Tea aroma is formed by a mixture of many volatile aromatic components with different contents and properties. With identification, there are about 700 kinds of aromatic substances in tea, but the main components are only dozens of them. Among them, there're about 100 kinds of components identified in fresh tea leaf, and more than 200 and 400 kinds for green and black tea respectively. With the continuous development of analysis and detection technologies, new aroma substances are still being discovered.

Chemical composition of tea aroma—According to the analysis of modern science and technology, the composition of aromatic substances in tea can be divided into 15 categories. They are classified into hydrocarbons, alcohols, aldehydes, acids, esters, ketones, esters, lactones, phenols, peroxides, sulfur-containing compounds, pyridines, pyrazines, quinolines, aromatic amines and so on.

Besides the aromatic substances, some other tea components could also have certain fragrance-enhancing effects. For example, some amino acids, such as glutamic acid, alanine and phenylalanine would bring certain floral fragrance. Some sugar components behave a sweetly fragrant aroma. And if sugar and pectin caramelized to certain extent, they would show a caramel aroma.

Main factors affecting tea aroma—Tea from different origins and of different kinds could have their own unique aroma. The composition of tea aroma is very complex, and the formation of tea aroma is influenced by various factors. The diversification of tea aroma is mainly related to ① the tea varieties, ② cultivation condition, ③ measures of planting management, ④picking standards, ⑤processing technology, and ⑥storage process.

Types of tea aroma—According to the characteristics of tea aroma, its types could be roughly classified as follows:

Pekoe flavor type, which reflects the fresh leaves with pekoe in raw materials, that emitted a pekoe aroma after the normal tea-making process.

Tender flavor type, which reflects the fresh and soft leaves, that emitted a tender fragrance after a reasonable tea-making process.

Flowery flavor type, which would be divided into clear flower flavor type (flowery like orchid, honeysuckle, gardenia jasminoides), and sweet flower flavor type (such as flowery and honey flavor, osmanthus fragrance and rose fragrance).

Fruity flavor type, such as peach, snow pear, longan, apple and so on.

Clean aroma type, which includes clean and high, clean and fresh, clean and lasting, clean and pure, and so on.

Sweet aroma type, which includes sweet fragrance, sweet flower fragrance, dried fruit fragrance, sweet jujube fragrance, honey fragrance and so on.

Stale flavor type, which reflects the mature tea materials, that emitted the aroma after a post-fermentation and aging process. It could refer to arohid flavor, arohid flavor, betel nut aroma, jujube fragrance, herbal fragrance, and so on.

Pine smoky flavor type, which reflects a tea drying process including smoking with pine and cypress.

4. Tea taste formation

As a beverage, the value of tea is mainly focused on the flavored composition and beneficial substances dissolved in tea liquor. Therefore, the taste of tea liquor is the main item to make up the tea quality.

Chemical composition of tea taste—The chemical composition of tea taste is complex, and the comprehensive reaction of taste organs to these intricate flavored constituents provides a variety of tea liquor taste. Different kinds of, grades of and quality of tea differ in taste quality. It also differs due to a large difference in the type, content and proportion of flavored substances in tea.

In tea the main taste ingredients can be roughly divided into the following categories. They are irritating astringent taste substances, bitter taste substances, fresh taste substances, sweet taste substances and sour taste substances.

For astringent taste substances, polyphenols are the main ingredients, among which the ester catechins behave strong bitter and astringent taste, strong constringent characteristics and are the main parts for astringent taste. The non-ester catechins will bring slight astringent taste and refreshing aftertaste, and the flavonoids taste bitter and astringent, but astringency weakened after the automatic oxidation.

For bitter taste substances, the alkaloids (caffeine), anthocyanins, and tea saponins make up the main ingredients, while catechins and flavonoids and so on are both astringent and bitter.

Fresh and refreshing taste is of great significance in the evaluation of tea taste quality. Such substances mainly include free amino acids and the complexes of caffeine with amino acids, theaflavins and catechins. In tea liquor, there are also soluble peptides and trace amounts of nucleoside acid, succinic acid and other fresh flavor components.

For sweet taste substances, they are mainly the soluble sugars and some amino acids, such fructose, glucose, sucrose, maltose, glycine, alanine, serine, and so on.

For sour taste substances, they are components originally from tea leaves or formed in tea processing, such as some amino acids, organic acids, ascorbic acid, gallic acids, non-polyphenolic pigments and so on.

Types of tea taste—The types of tea taste could be roughly classified as: strong type, strong and heavy type, heavy and mellow type, heavy and thick type, fresh and mellow type, fresh and mellow type, fresh and brisk type, aging and mellow type, sweet and mellow type, mellow type, mellow and thick type, mellow and brisk type, mellow and normal type, mellow and clean type, neutral type, and so on.

5. Tea cream formation

Tea cream is a cloudy or hazy appearance in tea and ready-to-drink tea products on cooling and related cream down is highly undesirable in the tea beverage industry. Especially troublesome in refrigerated teas, creaming gives adversely affects consumer quality.

Chemical composition of tea cream—As reported, chemical complexes formed by interactions among caffeine, polyphenolics, proteins, metal ions, carbohydrates, and/or other reactive agents in tea infusions are known as "tea cream". Not only for black tea cream compounds, including polyphenols, caffeine, protein, and metal compounds also play an important role in nonfermented tea creaming. More specifically, caffeine, gallocatechin (GC) and epigallocatechin gallate (EGCG) are the predominant compounds in green-tea cream but thearubigins (TRs), caffeine and theaflavins (TFs) in black-tea cream.

Physicochemical mechanism of the formation of tea cream—The formation of tea cream is governed by various molecular types of interactions, including polyphenol-caffeine complexation and polyphenol-polyphenol interactions. Herein, polyphenol-caffeine complexation is influenced by a number of gallate and hydroxyl groups of the polyphenols. As known for black tea cream, the formation is probably the result of hydrophobic interactions and/or hydrogen bonding between caffeine and functional groups on the causative chemical constituents. For example, hydroxyl groups of polyphenols, peptide groups of proteins, and the keto-amide group of caffeine could interact to form poorly soluble complexes in black tea infusions. Tea catechins are also considered to be the key components in tea cream, which interact with caffeine and protein to induce creaming.

Consequences of tea cream on tea quality—Tea cream is formed whilst hot aqueous tea infusion cools down. As reported, the amount of tea cream produced and its composition were influenced by extraction temperature and pH. The tea leaf/water ratio determined the amount of tea cream formed but did not affect the cream composition. Once formed, it affects not only physical attributes but also potentially biological activities and sensory attributes with altered levels of astringency, aroma, color, and taste.

The turbidity of ready-to-drink teas can be reduced by physical filtration methods; however, loss of bioactive polyphenolics and organoleptic properties are potential concerns. Thus far, no studies have shown complete removal of tea cream without loss of organoleptic traits or lowered concentrations of bioactive compounds.

词汇表

Pekoe flavor 毫香，是指白毫显露的幼嫩芽叶所具有的香气
heavy and thick 浓厚，指茶汤滋味入口浓，收敛性强，回味有黏稠感
heavy and strong 浓强
mellow and normal 醇正，指茶汤滋味浓度适当，正常无异味
stale and mellow 陈醇，指茶叶陈化后具有的愉悦的滋味，无霉味
spiral shape 卷曲如螺形，指茶叶条索卷紧后呈螺旋状，为碧螺春等卷曲形名优绿茶之造型
cream down 茶汤冷却后出现浅褐色或橙色乳状的浑浊现象

思考题

1. 将下面的句子译成英文。

（1）茶汤滋味的化学组成较为复杂。正是人们的味觉感官对这些错综复杂的呈味成分的综合反映，构成了各式各样的茶汤滋味。

（2）茶饮料、速溶茶、茶浓缩汁的加工及贮藏过程中形成的茶汤沉淀，不仅影响产品的外观品质，同时也造成产品风味品质的下降。

（3）随茶黄素和咖啡碱质量浓度的增高，其聚合后原溶液和贮藏液的透光率降低，贮藏液的沉淀量增加；并且当固定茶黄素和咖啡碱二者中任一物质的质量浓度时，随另外一种物质质量浓度的增加，上述变化趋势显著。

2. 将下面的句子译成中文。

（1）Tea color includes the color of dry tea leaves, tea liquor and remained tea leaves after infusion. They are the comprehensive reflection of a series of degradation and oxidative polymerization of the chemical composition of tea in tea making process.

(2) Tea from different origins and of different kinds could have their own unique aroma. The composition of tea aroma is very complex, and the formation of tea aroma is influenced by various factors.

(3) Chemical complexes formed by interactions among caffeine, polyphenolics, proteins, metal ions, carbohydrates, and/or other reactive agents in tea infusions are known as "tea cream".

参考文献

[1] SHI Z P. Tea Evaluation and Inspection[M]. Beijing: China Agriculture Press, 2010.

[2] GB/T 14487—2017. Tea vocabulary for sensory evaluation[S].

[3] LIN X, CHENG Z, ZHANG Y, et al. Comparative characterization of green tea and black tea cream: physicochemical and phytochemical nature[J]. Food Chemistry, 2015, 173: 432-440.

[4] JÖBSTL E, FAIRCLOUGH J P A, DAVIES A P, et al. Creaming in black tea[J]. Journal of Agricultural and Food Chemistry, 2005, 53(20): 7997-8002.

Lesson 4 Tea Evaluation and Inspection

1. Tea sensory evaluation

Tea Sensory evaluation is the process for professional reviews to comprehensively analyze and evaluate tea quality factors, such as dry appearance, liquor color, aroma, taste and infused leaves, by normal discrimination of vision, olfaction, taste and touch.

In order to ensure the accuracy of sensory evaluation of tea, the skilled tea rating staff should have a keen sensory review, and also to have good environmental conditions, equipment conditions and orderly methods for tea evaluation, such as corresponding provisions for tea utensils, water quality, tea water ratio, evaluation steps and relative methods.

China has published the national standard GB/T 18797—2012 for the General requirement of the tea sensory test room, which is equal to ISO 8589: 2007 for sensory analysis-general guidance for the design of test rooms. As only when objective conditions are unified, we can achieve the approximation of subjective knowledge. After over a century of tea trade exchanges and scientific communication, this special near-ancient tea quality evaluation method, has been recognized throughout the world.

1) Tea sampling methods

The sampling methods for refined tea and compressed tea should be in accordance with GB/T 8302—2013 "Tea—sampling", and sampling methods for primary tea are as followed.

Uniform stack sampling—Mix the batch of tea into piles, and then take samples from each part of the pile. Sampling points should not be less than eight.

On-the-spot sampling—Take a small sample from the top, middle, bottom, left and right parts of each sample and place it in a homogeneous plate, and check whether the quality of each sample is the same. If there are obvious differences among the top, middle, bottom, left and right parts of a single sample, the tea should be poured out, fully mixed, and then sampled again.

Random sampling method—The number of samples is randomly selected according to GB/T 8302—2013, and then to sample with on-the-spot sampling method.

After sampling with the above methods, the original obtained tea samples should be fully mixed. Two samples of 100-200 g individually should be obtained by using a sample splitter or by diagonal quartile method for evaluation. One is directly for sensory evaluation and the other should be retained in reserve.

2) Tea evaluation items

Tea quality factors—For tender tea and primary tea, the tea evaluation contents are composed by "five quality factors", which is the appearance of dry tea (including shape, tenderness, color, integrity and evenness), liquor color, aroma, taste and infused leaves.

For refined tea, the tea evaluation contents are composed by "eight quality factors", which is the shape of dry tea, dry color, crushing, purity, liquor color, aroma, taste and infused leaves.

Tea quality elements—*Dry tea appearance*: it includes the shape, tenderness, dry color, integrity and cleanliness of dry tea. Shape refers to the modeling, size, thickness, width, length and so on. Tenderness refers to the growth degree of raw materials. Dry color refers to the color and gloss of the product. Integrity refers to the integrity and breakage degree of the product. Cleanliness refers to the contents of tea stalks, tea slices and non-tea inclusions.

For compressed tea, such as tuo tea, brick tea, cake tea, etc., their quality elements contains the shape specifications, tightness, evenness, surface finish and dry color. If with internal and external differences, whether the layers are delaminated and whether the inner materials are exposed shall be assessed in addition. For Fiuzhuan tea, the quality of "golden flower", consisting of the size, glossiness and uniformity, shall also be assessed.

i. Tea liquor color: it refers to the color types and chrominance, brightness and turbidity, etc., of tea liquor.

ii. Tea aroma: it refers to the type, intensity, purity and persistence of tea aroma.

iii. Tea taste: it refers to the intensity, thickness, mellowness, purity and freshness, etc., of tea liquor.

iiii. Infused leaves: it refers to the tenderness, color, brightness and evenness (including evenness of tenderness and evenness of color) of infused leaves.

3) Methodology for sensory evaluation of tea.

Evaluation methods for dry tea appearance—After splitting into 100-200 g, put the representative tea samples in the appraisal tray. Hold the tray diagonally with both hands, and use the method of rotary sieve rotation, to make the tea samples stratified according to the order of thickness, length, size and brokenness, and closed in the middle of the appraisal tray in a round steamed bun-like shape. To evaluate the quality elements for dry tea appearance, use visual discrimination by observing the upper, middle and lower layer of tea samples, and inspect repeatedly.

Preparation methods for tea liquor—*Column Cup Method (standard cup-bowl method) for Black Tea, Green Tea, Yellow Tea, White Tea and Oolong Tea*: Put the representative tea samples of 3.0 or 5.0 g into standard column cup, and keep the ratio of tea to water ($m:V$) to be 1 : 50. The tea shall be filled with boiling water, covered and timed. The brewing time shall be applied according to Table 7.1, and tea liquor filtered out to bowls with fixed order and speed. The infused leaves shall be remained in the cup. The evaluation of intrinsic quality should be carried out in the order of liquor color, aroma, taste and remained leaves.

Table 7.1 Quantity and brewing time for different tea samples

Tea Types	Sample Quantity/g	Brewing Time/min
Black tea, green tea	3.0 or 5.0	5
Tender green tea	3.0 or 5.0	4
Oolong tea (strip or curl shape)	3.0 or 5.0	5
Oolong tea (round or granular shape)	3.0 or 5.0	6
White tea	3.0 or 5.0	5
Yellow tea	3.0 or 5.0	5

Covered Bowl (Gaiwan) Method for Oolong tea: Put the representative tea samples of 5.0 g into standard covered bowl, and keep the ratio of tea to water ($m:V$) to be 1 : 22. The tea shall be fully filled with boiling water, then covered and timed after the liquid foam removed with lid. After 1 minute, the fragrance of the inner side of lid is evaluated. At the 2 minutes, the liquor poured from covered bowl to tasting bowl, is evaluated by color and taste. For the subsequent second infusion, the inner side fragrance of lid is evaluated after 1 to 2 minute brewing, and the liquor color and tea taste are evaluated at 3 minute brewing. The continuous third infusion, the inner side fragrance of lid is evaluated after 2 to 3 minute brewing, and the liquor color and tea taste are evaluated at 5 minute brewing. Finally, sniff the infused leaves and evaluate remained leaves. The evaluation results shall be mainly based on the second infusion and take the first and third infusions into consideration as the general one.

Column Cup Evaluation Method for Dark tea (loose tea): Put the representative tea samples of 3.0 or 5.0 g into standard column cup, and keep the ratio of tea to water ($m : V$) to be 1 : 50. The tea shall be filled with boiling water, covered and timed. The first infusion by 2-minute brewing is to evaluate the liquor color, aroma and taste after leaching the liquor into bowl. The subsequent second infusion by 5-minute brewing is to evaluate the liquor color, aroma, taste and remained leaves after leaching the liquor into bowl. The liquor color evaluation results shall be based on the first infusion, and for aroma and taste based on the second infusion.

Column Cup Method for Compressed tea: Put the representative tea samples of 3.0 or 5.0 g into standard column cup, and keep the ratio of tea to water ($m : V$) to be 1 : 50. The tea shall be filled with boiling water, covered and timed. The first infusion by 2 to 5-minute brewing is to evaluate the liquor color, aroma and taste after leaching the liquor into bowl. The subsequent second infusion by 5 to 8-minute brewing is to evaluate the liquor color, aroma, taste and remained leaves after leaching the liquor into bowl. The evaluation results shall be mainly based on the second infusion and take the first infusions into consideration as the general one.

Column Cup Method for Scented tea: Remove the flower inclusions such as petals, calyx, pedicels and so on in tea samples, put the representative tea samples of 3.0 g into standard column cup, and keep the ratio of tea to water ($m : V$) to be 1 : 50. The tea shall be filled with boiling water, covered and timed.

The first infusion by 3-minute brewing is to evaluate the liquor color, aroma (freshness and purity) and taste after leaching the liquor into bowl. The subsequent second infusion by 5-minute brewing is to evaluate the liquor color, aroma (concentration and persistence), taste and remained leaves after leaching the liquor into bowl. The evaluation results shall take the first and second infusions into consideration as the general one.

Evaluation skills for intrinsic quality—

Tea liquor color: According to the liquor color quality elements, an attention should be paid to the influence of light and evaluation equipment for visual discrimination. The position of the standard bowl could be changed to reduce the influence of environmental light on the color of the tea liquor.

Tea aroma: With the cup in one hand and the lid in another hand, close to nostrils, to sniff the aroma by half opening of lid for 2 to 3 seconds, and repeat 1 to 2 times. According to the aroma quality elements, to evaluate the tea aroma with combination of hot sniffing (cup temperature around 75 ℃), warm sniffing (cup temperature around 45 ℃) and cold sniffing (room temperature).

Tea taste: Take a proper amount of tea liquor (about 5 mL) cooled naturally to around

50 ℃ with a teaspoon to suck it and circulate to make liquor touch all parts of tongue, spit out (preferred) or swallow it.

Infused leaves: A black wooden evaluation plate shall be used for refined tea, while a white enamel plate for primary tea, tender green tea and oolong tea. All leaves in the cup should be poured into the evaluation plate. If the white enamel plate is used, appropriate amount of water shall be added to make leaves floating. According to leaves quality elements, the infused leaves are evaluated by visual inspection and hand touching, etc.

2. Tea quality inspection

Tea quality inspection is to carry out physiochemical investigation for tea samples, including physical examination and chemical examination, only through which the tea samples could enter the market as their guarantee of the quality and safety.

1) Tea physical inspection

Tea physical inspection is a series of technical methods to detect and maintain tea quality by physical examination. In accordance with provisions of China's relevant standards, the tea physical inspections include the examination of tea dust, broken tea, tea package, foreign matter, quantitative inspection and so on. Other items not specified in the relevant standards, but reflecting the quality of tea, such as dry tea bulk weight, specific capacity test, tea soup colorimetric, may be applied.

Sampling for tea physical inspection: Sampling refers to the implementation of tea quality examination of a certain number of samples representing the quality of the whole batch of tea products in accordance with the provisions of standards, which is the basis for ensuring the authenticity of the inspection results. With the variation of tea categories, origin, seasons, processing and so on, tea quality is very different, thus it's an extremely meticulous work to make representative sampling.

In accordance with national standards, sampling should be taken in batches, no matter packaged in boxes or baskets. The quantity for tea sampling is stipulated as in Table 7.2. In general, 1 piece shall be taken if tea quantity in a batch is 1-5 pieces, 2 pieces shall be taken if 6-50 pieces in a batch. One additional sampling piece increased for each 50 pieces if samples more than 50 pieces, and one additional piece for each 100 pieces if more than 500 pieces, and one additional piece for each 500 pieces if more than 1000 pieces in a batch.

Physical inspection of dust and broken tea: It is inevitable to produce some dust and broken leaf in the primary and refined tea, the existence of which has a direct impact on the shape and the uniform of tea products. They may make liquor color dark and taste bitter after brewing. Therefore, the amount of dust and broken leaf shall be limited in tea product standards.

Table 7.2   Total tea amount and sampling quantity

Tea Amount	Sampling Quantity	Tea Amount	Sampling Quantity	Tea Amount	Sampling Quantity
1–5	1	351–400	9	1001–1200	17
6–50	2	401–450	10	1201–1400	18
51–100	3	451–500	11	1401–1500	19
101–150	4	501–600	12	1501–2000	20
151–200	5	601–700	13	2001–2500	21
201–250	6	701–800	14	2501–3000	22
251–300	7	801–900	15	3001–3500	23
301–350	8	901–1000	16	3501–4000	24

(Unit of quantity: pieces)

Physical inspection of tea stems: Tea stems are remnant tender stems and stalks in tea products after processing, mainly caused by materials rough picking or not neat sorting in the refining process. Tea stems, existed in all types of primary teas, and their size, length, color, etc., would affect the shape of tea and its intrinsic quality. Generally, for black tea and green tea, the inspection of tea stems should be in accordance with the Trade Standard Sample or transaction sample.

Physical inspection of non-tea matter: In the process of tea production, some kinds of non-tea matter may be entered in final tea products and have a safety risk. Among them, insect carcasses, iron chips, sand, plastic, broken glass slices may be existed. Those non-tea matter shall be strictly examined.

2) **Tea chemical inspection**

Tea chemical inspection is a series of technical methods to detect tea components to determine whether the product meets the quality requirement and drinking demands by chemical examination. They are divided into specific chemical inspection and general chemical inspection.

Specific chemical inspection: In addition to moisture inspection and ash inspection as statutory chemical inspection project, the specific designated chemical inspectionitems are mainly the amount of polyphenols, caffeine, water extract, crude fibre, pesticide residues and other items.

For moisture inspection, the method of drying in an oven at (103 ± 2) ℃ for around 4 hours to constant mass is applied by the requirements of tea standard ISO 1573 and GB 5009. 3—2016. The final loss of weight shall be calculated as the tea moisture content.

Total ash content of tea is not only the quality index of tea, but also the hygienic index. As the requirements of ISO 1575 and GB/T 8306—2013, the total ash content of tea could be examined by burning the samples at (525 ± 25) ℃ or 700 ℃ to the constant mass. Moreover,

Tea ash could be divided into water-soluble ash and acid-soluble ash.

The amount of water extracts in tea was positively correlated with the quality, and closely related to the tenderness of fresh leaves, as well as tea varieties, cultivation conditions and processing techniques. The method shall be in accordance to ISO 9768 and GB/T 8305—2013.

For total tea polyphenols inspection, it is determined by a colourimetric assay using Folin-Ciocalteu phenol methods as the requirements of ISO 14502 and GB/T 8313—2018. For caffeine inspection, high performance liquid chromatography (HPLC) or ultraviolet spectrophotometry is applied in accordance with ISO 10727 and GB/T 8312—2013.

Here are chemical requirements for green tea in ISO 11287 "green tea—definition and basic requirements" as shown in Table 7.3.

Table 7.3　　　　Total tea amount and sampling quantity

Characteristics	Requirement	Test method
Water extract, % mass fraction	32 min	ISO 9768
Total ash, % mass fraction	8 max / 4 min	ISO 1575
Water-soluble ash, % mass fraction of total ash	45 min	ISO 1576
Alkalinity of water-soluble ash (as KOH), % mass fraction	1.0 min[a] / 3.0 max[a]	ISO 1578
Acid-insoluble ash, % mass fraction	1.0 max	ISO 1577
Crude fibre, % mass fraction	16.5 max	ISO 5498 or ISO 15598[b]
Total catechins, % mass fraction	7 min	ISO 14502-2
Total polyphenols, % mass fraction	11 min	ISO 14502-1
Ratio total catechins to total polyphenols, mass fraction	0.5 min	

Note: Green tea specifically cultivated in a manner that suppresses the catechin and total polyphenol content, including Tencha (Matcha) and Gyokuro, has minimum 8% mass fraction total polyphenols and minimum 5% mass fraction total catechins.

[a] When the alkalinity of water-soluble ash is expressed in terms of millimoles of KOH per 100 g of ground sample, the limits shall be: 17.8 min; 53.6 max.

[b] The specific method for the determination of crude fibre in tea is specified in ISO 15598; however, for the purpose of routine estimation, the general method specified in ISO 5498 is adequate. In cases of dispute, the method of determination should always be that specified in ISO 15598. The requirement remains unchanged, regardless of the method used.

General chemical and microbial inspection: General chemical inspection includes the examination of theaflavins, thearubins, crude fibers, radioactive substances, trace elements and other chemical tests. Among them, HPLC or spectrophotometer detection method could be used for theaflavins and thearubigin, and gravimetric method shall be used for crude fibers.

In addition, it is also very important to inspect pesticide residues, heavy metals, microbes and mycotoxins in tea. All of those items shall be met to the compulsory requirements of National food safety standards. The exporting teas should comply with the requirements of the designated importing countries.

> ## 词汇表
>
> Tea sensory evaluation 茶叶感官审评，审评人员运用正常的视觉、嗅觉、味觉、触觉等辨别能力，对茶叶产品的外形、汤色、香气、滋味与叶底等品质因子进行综合分析和评价的过程
> Uniform stack sampling method 匀堆取样法，指将该批茶叶拌匀成堆，然后从堆的各个部位分别扦取样茶的方式
> On-the-spot sampling method 就件取样法，从每件上、中、下、左、右五个部位各扦取一把小样置于扦样匾盘中，并查看样品间品质是否一致。若单件的上、中、下、左、右五部分样品差异明显，应将该件茶叶倒出，充分拌匀后，再扦取样品
> primary tea 初制茶，也称毛茶（crude tea），指茶鲜叶加工后还需要再精细加工的产品
> refined tea processing 精制茶加工，指对毛茶或半成品原料茶进行筛分、轧切、风选、干燥、匀堆、拼配等精制加工茶叶的生产
> moisture content 含水量
> water extract 水浸出物
> % mass fraction 质量分数（%）
> total ash 总灰分
> alkalinity of water-soluble ash（as KOH）水溶性灰分碱度
> crude fibre 粗纤维

思考题

1. 将下面的句子译成英文。

（1）对照一组标准样品，比较未知茶样品与标准样品之间某一级别在外形和内质的相符程度。

（2）审评过程中由主评先评出分数，其他人员根据品质标准对主评出具的分数进行修改与确认，对观点差异较大的茶进行讨论，最后共同确定分数，如有争论，投票决定。

（3）根据审评知识与品质标准，按外形、汤色、香气、滋味和叶底"五因子"，采用百分制，在公平、公正条件下给每个茶样每项因子进行评分。

2. 将下面的句子译成中文。

（1）For tender tea and primary tea, the tea evaluation contents are composed by "five quality factors", which is the appearance of dry tea（including shape, tenderness, color, integrity and evenness）, liquor color, aroma, taste and infused leaves.

（2）The tea aroma quality is evaluated by combination of hot, warm and cold sniffing.

（3）Tea quality inspection is to carry out physiochemical investigation for tea samples, including physical and chemical examination, only through which the tea samples could enter the market as their guarantee of the quality and safety.

参考文献

[1] SHI Z. P. Tea Evaluation and Inspection [M]. Beijing: China Agriculture Press, 2010.

[2] ZHANG Y, LIU X, LU C. Study on primitive morpheme in sensory terminology and flavor wheel construction of Chines tea [J]. Journal of Tea Science, 2019, 39(4). DOI:10.13305/j.cnki.jts.20190530.001.

[3] GB/T 23776—2018 Methodology for sensory evaluation of tea [S].

[4] GB/T 14487—2017. Tea vocabulary for sensory evaluation [S].

[5] GB/T 18797—2012. General requirement of the tea sensory test room [S].

[6] GB 2762—2017. National food safety standard—Maximum levels of pollutants in food [S].

[7] GB 2763—2016. National food safety standard—Maximum residue limits for pesticides in food [S].

[8] GB 2763.1—2018. National food safety standard—Maximum residue limits for 43 pesticides in food [S].

[9] ISO 11287—2011. green tea—definition and basic requirements [S].

UNIT Eight　Tea and Health Effects

Lesson 1　Antioxidant and Pro-oxidant Effects

1. Redox chemistry of tea polyphenols

1) Direct antioxidant effects

The antioxidant activity of (-)-epicatechin (EC), (-)-epigallocatechin (EGC), (-)-epicatechin-3-gallate (ECG), and EGCG has been demonstrated in a number of *in vitro* and chemical-based assays. The chemistry underlying activity results mainly from hydrogen atom transfer (HAT) or single electron transfer reactions (SET), or both, involving hydroxyl groups. These groups are constituents of the B-rings of EC and EGC, and both B-and D-rings of ECG and EGCG.

As chain-breaking antioxidants, tea catechins are thought to interrupt deleterious oxidation reactions by HAT mechanisms, the most important being lipid peroxidation:

$$L_1H \rightarrow L_1\cdot \text{ (initiation)} \tag{1}$$

$$L_1\cdot + O_2 \rightarrow L_1O_2\cdot \text{ (formation of peroxyl radical, } \sim 10^9 M^{-1}s^{-1}) \tag{2}$$

$$L_1O_2\cdot + L_2H \rightarrow L_1OOH + L_2\cdot \text{ (chain propagation, } \sim 10^1 M^{-1}s^{-1}) \tag{3}$$

Lipid peroxidation is a radical chain reaction in which hydrogen atoms are abstracted [Rxn. (1), symbol for drug reaction] from unsaturated fatty acids. In the absence of chain-breaking antioxidants, these radicals resulting in new lipid alkyl radicals ($L_2\cdot$), thus propagating the chain reaction. Fortunately, the reaction between lipid peroxyl radicals and unoxidized lipids [Rxn. (3)] is relatively slow (ca. $10^1 M^{-1} s^{-1}$), affording phenolic antioxidants (PhOH) the opportunity to intercept peroxyl radicals and interrupting chain propagation.

$$L_1O_2\cdot + PhOH \rightarrow L_1OOH + PhO\cdot \text{ (chain interruption, } k_4 > k_3) \tag{4}$$

Tea catechins are also known to quench free radicals by SET reactions, wherein phenolic radical cations are first formed followed by deprotonation:

$$PhOH + LO_2\cdot \rightarrow PhOH^+ + LO_2^- \tag{5}$$

$$PhOH^+ + H_2O \longleftrightarrow PhO\cdot + H_3O^+ \tag{6}$$

$$LO_2\cdot + H_3O^+ \longleftrightarrow LOOH + H_2O \tag{7}$$

Taken together, the chain-breaking activity of tea catechins are known to involve both HAT and SET mechanisms, it appears HAT mechanisms dominate and that this activity is restricted to the B-ring, even in the case of catechins with galloyl moieties comprising their D-rings (ECG, EGCG). The radical scavenging activity of the major tea catechins on peroxyl radicals has been shown to follow the order: EC < ECG ≈ EGC < EGCG.

2) Trace metal catalysis of polyphenol oxidation

The rate of catechin oxidation increases as a function of increasing pH. The base-catalyzed oxidation of phenolic compounds is often referred to as "auto-oxidative" because it is thought that oxygen reacts faster with phenolate anions. Transition metals (e.g., iron and copper) are capable of initiating phenolic oxidation and are essential catalysts in this process. This reaction also yields a reactive oxygen species, namely superoxide ($O_2^-\cdot$) or its protonated form, the hydroperoxyl radical ($HO_2\cdot$), under acidic conditions [Rxn. (9)], that is further reduced to hydrogen peroxide [Rxn. (11)]:

$$PhOH + M^{n+} \rightarrow PhO\cdot + M^{(n-1)+} \tag{8}$$

$$M^{(n-1)+} + O_2 \rightarrow M^{(n-1)+} + O_2^-\cdot \tag{9}$$

$$PhO\cdot + O_2 \rightarrow QPh + O_2^-\cdot \tag{10}$$

$$O_2^-\cdot + PhOH \rightarrow PhO\cdot + H_2O_2 \tag{11}$$

Semiquinone radicals and, eventually, quinones are generated in the process, which are highly electrophilic species that can react with free thiol-bearing compounds to form stable conjugates. However, the reduced forms of transition metals are known to catalyze lipid hydroperoxide and hydrogen peroxide decomposition to lipid alkoxyl (LO·) and hydroxyl radicals (HO·), respectively. These results suggest a role for transition metals in some of the pro-oxidant effects of green tea polyphenols.

The many elegant chemistry experiments that have been conducted on the redox effects of tea polyphenols provide some potential mechanisms by which to understand the antioxidant and pro-oxidant effects of these compounds in biological systems (Figure 8.1). Nonetheless, further investigations are

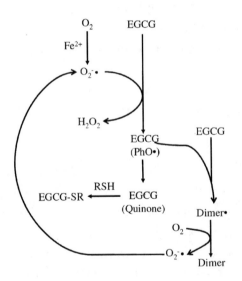

Figure 8.1 Oxidative reaction between EGCG, superoxide and ferric iron resulting in the production of oxidative stress, EGCG dimers, and EGCG - cysteine conjugates (EGCG-SR)

PhO· = semiquinone radical

needed.

2. Role of antioxidant effects in the cancer preventive activity of green tea polyphenols

1) Direct antioxidant effects

As described above, tea polyphenols are strong radical scavengers and metal chelators in model chemical systems, and these effects correlate with the presence of the dihydroxy and trihydroxy groups. An increasing number of studies have also demonstrated these antioxidative effects *in vivo*. In the study by Hakim et al., supplementation of heavy smokers (>10 cigarettes per day) with 4 cups of decaffeinated green tea (73.5 mg catechins per cup) per day for 4 months reduced urinary 8-OHdG levels by 31% compared to control.

2) Induction of endogenous antioxidants

Studies have shown that tea treatment can induce Phase II metabolism and antioxidant enzymes in both animal models and humans. A study in China of subjects in high risk areas for liver cancer found that supplementation with green tea polyphenols (500 or 1,000 mg/day) resulted in increased urinary excretion of aflatoxin B_1 (AFB_1) mercapturic acid conjugates (AFB_1-NAC) compared to baseline.

3. Role of pro-oxidant effects in the cancer preventive activity of green tea polyphenols

1) Generation of reactive oxygen species by tea polyphenols

Under typical cell culture conditions, green tea polyphenols are unstable and undergo auto-oxidative reactions resulting in the production of ROS. Incubation of HL-60 human leukemia cells with tea catechins in the presence of Cu(II) resulted in the formation of 8-oxo-guanosine.

2) Induction of apoptosis and inhibition of cell proliferation

Yang et al., have reported that production of ROS plays an important role in the pro-apoptotic effects of EGCG and other tea polyphenols against cancer cell lines. EGCG-mediated apoptosis was reduced by approximately 50% by inclusion of exogenous catalase, whereas inclusion of catalase had no effect on the pro-apoptotic effects of TFdiG. These results suggested that TFdiG can induce apoptosis by an ROS independent mechanism in this model system.

3) Effects on cell signaling

Several studies have shown that the effect of EGCG and other tea polyphenols on key cell signaling pathways may be dependent on polyphenol-mediated reactive oxygen species. At least, transforming growth factor (TGF)β signaling and epidermal growth factor (EGF) mediated signaling were involved.

4) Pro-oxidant effects in animal models

Following treatment of CF-1 mice with 400 mg/kg, i.p. EGCG, EGCG-2′-cysteine and EGCG-2″-cysteine could be detected in the urine. These metabolites are hypothesized to arise from the reaction of EGCG quinone intermediates with the thiol moiety of cysteine, and suggests that at high doses EGCG may have pro-oxidant effects *in vivo*.

4. Conclusions

It is possible that the direct antioxidant effects of tea polyphenols may play an important role under certain circumstances. There is considerable and increasing evidence that green tea and EGCG can enhance the expression of endogenous antioxidant systems in both animal models and human subjects. These effects include modulation of plasma and tissue levels of SOD, catalase, and enzymes related to glutathione metabolism.

Although tea polyphenols have generally been regarded as antioxidants, the emerging evidence for the pro-oxidant effects of these compounds is interesting. One particularly interesting possibility is that tea polyphenols, at dietary levels, induce low levels of oxidative stress which "prime" endogenous antioxidant systems to deal with larger oxidative insults arising from ultraviolet light and other carcinogens (Figure 8.2).

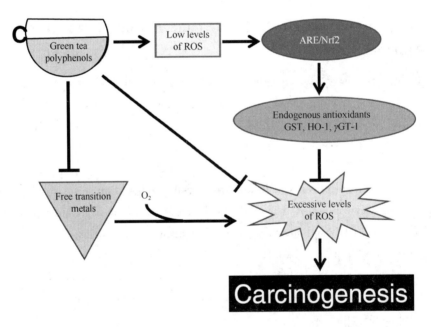

Figure 8.2 Proposed model for the mechanism by which the antioxidant and/or pro-antioxidant effects tea polyphenols inhibit the carcinogenic process

ROS = reactive oxygen species, ARE = antioxidant response element, GST = glutathione Stransferase, γGT = gamma glutamyltransferase

词汇表

pro-oxidant [pˈrəʊɒksˈɪdænt] n. 促氧化剂

redox [ˈredɒks] n. 氧化还原作用

in vitro [ɪn ˈviːtrəʊ] 在生物体外；离体

in vivo [ɪn ˈviːvəʊ] 在活的有机体内；活体内

peroxidation [pəˌrɒksɪˈdeɪʃən] n. 过氧化反应

Rxn., symbol for drug reaction = reaction 化学反应

alkyl radical [ˈælkɪl ˈrædɪkəl] 烷自由基

hydroperoxyl radical 过氧羟自由基

hydroperoxide 氢过氧化物

Fenton reagent (reaction) 芬顿试剂或反应。是指由过氧化氢和亚铁离子组成的具有强氧化性的体系，可将很多已知的有机化合物如羧酸、醇、酯类氧化为无机态

alkoxyl radical [ˈɔːlkɒksɪl] 烷氧基

bond dissociation enthalpy [enˈθælpɪ] 键离解能

semiquinone [semɪkwɪˈnəʊn] n. 半醌

hypochlorous acid 次氯酸

peroxynitrous acid 过氧亚硝酸

Data Base Design Evaluator (DBDE) 数据库设计评价程序

bio-availability [ˌbaɪəʊəˌveɪləˈbɪlɪtɪ] n. (药物或营养素的)生物药效率，生物利用度。是反映所给药物进入人体循环的药量比例，它描述口服药物由胃肠道吸收，及经过肝脏而到达体循环血液中的药量占口服剂量的百分比。包括生物利用程度与生物利用速度

base-catalyzed 碱催化

anion [ˈænaɪən] n. 阴离子；对应的为阳离子 cation [ˈkætaɪən]

thermodynamical [θɜːməʊdaɪˈnæmɪkəl] adj. 热力学的

quantum [ˈkwɒntəm] n. [物] 量子

Pauli Exclusion Principle [化] 泡利[不相容]原理，是微观粒子运动的基本规律之一

catalyst [ˈkætəlɪst] n. 催化剂

reactive oxygen species (ROS) 活性氧，是体内一类氧的单电子还原产物，包括氧的一电子还原产物超氧阴离子 O_2^-、二电子还原产物 H_2O_2、三电子还原产物羟基自由基 •OH 以及一氧化氮 NO 等。线粒体是活性氧的一个重要来源。在大多数细胞中超过 90% 的氧是在线粒体中消耗的，其中 2% 的氧在线粒体内膜和基质中被转变成为氧自由基。体内正常代谢可以产生活性氧，但若活性氧过剩会对细胞造成伤害，甚至导致细胞死亡，诱发各种心血管疾病、神经性疾病及肿瘤等 200 多种疾病，还可使组织器官衰老

protonate [ˈprəʊtəneɪt] vt. 使质子化，给…加质子

desferrioxamine [diːsferɪˈɒksəmaɪn] n. 去铁胺(敏)。为铁的络合剂，与三价铁络合成无毒物排出，为铁中毒的解毒剂

diethylenetriaminepentaacetic acid (DTPA) 二乙烯三胺五乙酸

catalase [ˈkætəleɪs] n. 过氧化氢酶

superoxide dismutase (SOD) 超氧(化)物歧化酶

catecholate-and gallate-ferric complexes 儿茶酚酸和没食酸亚铁复合物

cell line 细胞系

animal model 动物模型

EGCG-cysteine conjugates（EGCG-SR）EGCG 半胱氨酸共轭物(通过硫键)

radical scavenger 自由基清除剂

metal chelator 金属络合剂

Phase Ⅱ metabolism Ⅱ相代谢。多数药物在肝脏内代谢分为两个阶段：I相反应和Ⅱ相反应。第一阶段为氧化、还原及水解、去甲基反应，产生一系列肝细胞毒性产物，主要包括亲电子基和氧自由基，也称作I相反应。第二阶段为结合反应，即药物解毒过程，也称作Ⅱ相反应，如与葡萄糖醛酸、硫酸盐、氨基酸等结合，进一步使药物分子的极性增大，有利于从尿中排出

8-oxo-guanosine 8-氧鸟苷。人体尿液中 RNA（核糖核酸）氧化代谢产物 8-氧化鸟苷（8-oxoGsn）的含量可以作为人体新的衰老标志物

apoptosis ［细胞］凋亡。指为维持内环境稳定，由基因控制的细胞自主的有序的死亡。细胞凋亡与细胞坏死(necrosis)不同，细胞凋亡不是一件被动的过程，而是主动过程，它涉及一系列基因的激活、表达以及调控等的作用，它并不是病理条件下，自体损伤的一种现象，而是为更好地适应生存环境而主动争取的一种死亡过程

pro-apoptotic effect 促凋亡效应

human bronchial epithelial cell 人支气管上皮细胞

cell signaling 细胞信号传导。信号转导通常包括以下步骤：特定的细胞释放信息物质或吸收的外源物质→信息物质经扩散或血循环到达靶细胞→与靶细胞的受体特异性结合→受体对信号进行转换并启动细胞内信使系统→靶细胞产生生物学效应。通过这一系列的过程，生物体对外界刺激作出反应

transforming growth factor（TGF）β 转化生长因子-β，对细胞的生长、分化和免疫功能都有重要的调节作用

epidermal growth factor（EGF）表皮细胞生长因子，可刺激多种细胞的增殖，主要是表皮细胞、内皮细胞

intermediate ［ˌɪntəˈmiːdiət］ n. 中间物，中间分子；adj. 中间的，中级的

thiol ［ˈθaɪəʊl］ n. 硫醇；巯基

cysteine ［ˈsɪstɪn］ n. 半胱氨酸

carcinogenesis ［kɑːsɪnəʊˈdʒenɪsɪs］ n. 癌病变，致癌作用

endogenous antioxidant 内源性抗氧化剂

oxidative stress 氧化应激，是指体内氧化与抗氧化作用失衡，倾向于氧化。氧化应激是自由基在体内产生的一种负面作用，并被认为是导致衰老和疾病的一个重要因素

chemoprevention 化学预防（癌症），是指对无症状的使用药物、营养素(包括无机盐)、生物制剂或其他天然物质作为一级、二级预防为主的措施，提高人群抵抗疾病的能力，以防止某些疾病

endogenous antioxidant systems 内源性抗氧化系统，包括酶抗氧化系统和非酶抗氧化系统。酶抗氧化系统主要由超氧化物歧化酶(SOD)、过氧化氢酶(CAT)、谷胱甘肽过氧化物酶(GSH-Px)和谷胱甘肽还原酶(GR)和醛酮还原酶(AR)等抗氧化酶组成；非酶抗氧化系统主要由各种非酶抗氧化剂组成：①脂溶性抗氧化剂，如维生素 E、类胡萝卜素(CAR)、辅酶(CoQ)等；②水溶性抗氧化剂，如维生素 C、谷胱甘肽(GSH)、硫辛酸等；③硒、铜、锌、锰等微量元素；④人体代谢产物，如尿酸盐等、内源性褪黑激素(MT)等；⑤植物化学物质，如植物纤维、

> 植物多糖、植物甾醇、酚类化合物、有机硫化合物、萜类化合物、天然色素和部分中草药成分等。它们大部分通过体外摄入
>
> glutathione S transferase（GST）谷胱甘肽 S-转移酶，是一组与肝脏解毒功能有关的酶。该酶主要存在于肝脏内。当肝细胞损害时，酶迅速释放入血，导致血清 GST 活力升高
>
> HO-1（Heme Oxygenase-1）血红素氧合酶-1 或血红素加氧酶-1(EC 1.14.99.3)，属氧应激诱导型，是血红素分解代谢过程中的重要酶，将血红素转化为胆绿素
>
> gamma glutamyltransferase（γGT-1 or GGT）γ-谷氨酰基转移酶-1，主要存在肝细胞膜和微粒体上，参与谷胱甘肽的代谢

思考题

1. 将下面的句子译成英文。

（1）绿茶中含有大量的抗氧化物质，如儿茶素、黄酮醇、酚酸、维生素 C、维生素 E 和胡萝卜素等。其中 EGCG 是最主要的抗氧化物质。

（2）儿茶素类物质自身具有优异的抗氧化活性，它们不仅具有直接的抗氧化作用，而且还可以通过提高体内内源酶的活性起到间接的抗氧化作用。

（3）在特定条件下，儿茶素类可以与过渡金属离子如 Fe^{3+}、Cu^{2+} 等反应，产生活性氧，诱导癌细胞凋亡。

2. 将下面的句子译成中文。

（1）Several studies have shown that the effect of EGCG and other tea polyphenols on key cell signaling pathways may be dependent on polyphenol-mediated reactive oxygen species.

（2）Alternatively, it may be that the bio–availability and tissue distribution of tea polyphenols is limited to such an extent that the antioxidative effects observed *in vitro* are very unlikely or impossible *in vivo*.

（3）One particularly interesting possibility is that tea polyphenols, at dietary levels, induce low levels of oxidative stress which "prime" endogenous antioxidant systems to deal with larger oxidative insults arising from ultraviolet light and other carcinogens.

参考文献

[1] LAMBERT J D, ELIAS R J. The antioxidant and pro-oxidant activities of green tea polyphenols: A role in cancer prevention[J]. Achieves of Biochemistry and Biophysics, 2010, 501: 65-72.

[2] DAS K, BHATTACHARYYA J. Antioxidant functions of green and black tea[M]// PREEDY V R. Tea in health and disease prevention. San Diego, U.S.: Academic Press, 2013: 521-527.

[3] HALLIWELL. Are polyphenols antioxidants or pro-oxidants? What do we learn from cell culture and *in vivo* studies? [J]. Archives of Biochemistry and Biophysics, 2008, 476:

107-112.

[4] ROOS B D, DUTHIE G G. Role of dietary pro-oxidants in the maintenance of health and resilience to oxidative stress[J]. Mol Nutri Food Res, 2015, 59: 1229-1248.

[5] FERNANDO W, RUPASINGHE V, HOSKIN D W. Dietary phytochemicals with anti-oxidant and pro-oxidant activities: A double-edged sword in relation to adjuvant chemotherapy and radiotherapy?[J]. Cancer Letters, 2019, 452: 168-177.

Lesson 2 Anticancer Effects

1. Cancer prevention by tea: The evidence from cells and animal studies

Cancer prevention by tea and its components has been evidenced in various cells and animal models, and the mechanisms actions have been widely investigated. The anticancer effects of tea will be discussed given the mechanisms of actions.

1) Inhibition on cell proliferation

Various tea and its components possessed cancer cell-selective toxicity via inhibiting proliferation-related signaling pathways, and subsequently preventing cancer formation. Epicatechin-3-gallate (EGCG) could inhibit the proliferation and growth of bladder cancer cells via downregulating the expression of nuclear factor-kappa B (NF-κB) in both protein and mRNA levels.

Tea has been demonstrated to regulate cell cycle related proteins, and induce cell cycle arrest to prevent carcinogenesis. EGCG treatment could upregulate the expression of B-cell translocation gene 2 and decrease cell cycle-related proteins, i.e. cyclins A, D and E, which induced cell cycle arrest at G1 phase in oral cell carcinoma.

2) Anti-angiogenesis

Angiogenesis is important for the growth of cancer because cancer cells could obtain more nutrients. Thus, anti-angiogenesis is essential to manage cancer. The prodrug of EGCG could inhibit the secretion of vascular endothelial growth factor A in endometrial cancer cells, thus suppressing cancer angiogenesis. Theaflavin-3,3′-digallate of black tea was effective in inhibiting angiogenesis by inactivating Akt/mTOR/p70S6 kinase/eukaryotic initiation factor 4E-binding protein-1 pathway and Akt/cMyc pathway in human ovarian carcinoma cells.

3) Induction of apoptosis

Apoptosis is closely associated with survival and proliferation of cancer cells. Thus, induction of apoptosis is important for management of cancer. Catechin extract was reported to induce apoptosis in prostate cancer cells by decreasing B-cell lymphoma 2 (Bcl-2) expression and increasing cytochrome c expression, triggering the activation of caspase-3. EGCG encapsulated in chitosan-based nanoparticles induced the cleavage of poly polymerases, increased the expression of pro-apoptotic protein Bcl-2-associated X protein and decreased the expression of anti-apoptotic protein Bcl-2.

4) Suppression on cell metastasis

The *in vivo* and *in vitro* studies revealed that green tea extract decreased the cancer metastasis. Green tea polyphenols restricted the invasion and promotion of breast cancer by inducing the tissue inhibitor of matrix metalloproteinase (MMP)-3, and inhibiting the invasiveness and gelatinolytic activity of matrix metalloproteinase-2 and MMP-9 which are major mediators in migration and invasion. Treatment with EGCG reduced the migration and invasive properties in nasopharyngeal carcinoma cells and induced downregulation of MMP-2 and MMP-9, which were associated with the suppression of extracellular signal-regulated kinase phosphorylation and activator protein 1 (AP-1) and Sp1 transactivation. Additionally, tea polysaccharide reduced the expression levels of MMP-2 and MMP-9, inhibiting the invasion of the mouse colon cancer cells.

5) Inhibition on cancer stem cells

Cancer stem cells (CSCs) have recently been considered playing key roles in the initiation, unlimited growth, recurrence, and metastasis of cancers. High levels of the stemness markers and EMT markers have been found in CSCs. Whereas, EGCG showed inhibitory effects on human CSCs by inhibiting the expressions of genes related to stemness markers and epithelial-mesenchymal transition (EMT) phenotypes. Additionally, EGCG was effective in repressing self-renewal and the expression of pluripotency-maintaining transcription factors in human CSCs.

6) Modulation on gut microbiota

The interplay between tea phytochemicals and gut microbiota could improve the anticancer effects of tea (Figure 8.3). A study showed that green tea consumption could elevate levels of short-chain fatty acid-producing bacterial genera, and reduce bacterial lipopolysaccharide synthesis, and decrease the functional pathways abundance relevance to cancer, which may ameliorate the inflammation and inhibit colorectal cancer.

7) Adjuvant therapy

Tea can enhance the efficacy of anticancer drugs, and have synergistic effects with other compounds against several cancers, which make tea an appropriate adjuvant. A study found that EGCG could enhance the anticancer efficacy of paclitaxel via promoting apoptosis and inhibiting invasion. Moreover, EGCG was demonstrated to enhance the efficacy of herceptin, an anticancer protein, through increasing its cancer selectivity, inhibiting cancer growth as well as prolonging the half-life of herceptin in the blood. Also, the anticancer effects of doxorubicin were potentiated by EGCG, which could improve its stability in blood circulation and enhance the permeability and retention effect of doxorubicin.

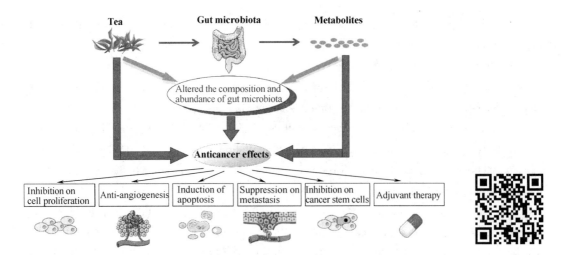

Figure 8.3 The crosstalk between tea, its metabolites and the altered gut microbiota plays an important role in improving the anticancer effects of tea

2. Epidemiological studies

The association between tea consumption and the risk of cancers has been widely investigated by epidemiological methods. A cohort study demonstrated that the increased tea intake was associated with lower hepatocellular cancer risk (HR, 0.41; 95% CI, 0.22-0.78), and nine prospective cohort studies of China, Japan and Singapore observed that the consumption of green tea was effective in reducing the risk of liver cancer in women (RR, 0.78; 95% CI, 0.64-0.96). Moreover, a meta-analysis revealed that the consumption of green tea was inversely associated with the risk of ovarian cancer (RR, 0.64; 95% CI, 0.45-0.90). However, the inconsistent results are also reported about effects of tea on several cancers in epidemiological studies. For example, a cohort study showed that the relationship between drinking tea and gastric cancer risk was non-significant. The inconsistent results from some epidemiological studies could be due to variables, such as the inaccurate assessment of tea usage, the lack of a standard for tea preparation, differences in the composition of teas derived from different areas, and genetic variability among subjects.

3. Clinical trials

Epidemiological and experimental studies have suggested a protective role of tea against several cancers, and some clinical trials have been also carried out to clarify and further evaluate effects of tea on cancers in human. A randomized controlled trial (RCT) observed that the intake of green tea extract containing 1,315 mg total catechins for one year reduced the percent mammographic density in young women and decreased the risk of breast cancer. Moreover, an RCT has studied the anticancer efficacy of brewed green tea in patients with prostate cancer (six cups daily), and the results showed that brewed green tea could decrease markers of cancer development and progression (NF-κB and urinary 8-hydroxydeoxyguanosine). Collectively, the clinical trials have demonstrated that tea and its phytochemicals could effectively prevent and manage several cancers.

4. Conclusions

Tea has shown preventive efficacy for several cancers, such as liver, lung, breast, prostate and colon cancers. Both cell and animal studies have demonstrated that tea exhibited anticancer efficacy by targeting molecular events relevant to cancer initiation and progression, with the inhibition on cell proliferation, anti-angiogenesis, induction of apoptosis and suppression on cell metastasis. CSCs could be effectively inhibited and the gut microbiota could be altered by tea, which was considered as the novel target of the anticancer mechanisms. Moreover, tea could enhance the efficacy of anticancer therapies. In addition, the epidemiological studies have revealed that the tea consumption has inverse association with the risk of several cancers, though there are still some inconsistent results. Clinical trials have found that intake of tea could protect against several cancers, further verifying the preventive role of tea in human.

词汇表

proliferation [prəˌlɪfəˈreɪʃn] n. 增殖
downregulate/ upregulate 下调/上调
carcinogenesis [kɑːsɪnəʊˈdʒenɪsɪs] n. 癌病变，致癌作用
nuclear factor-kappa B (NF-κB) 核因子 κB 蛋白。NF-κB 的错误调节会引发自身免疫病、慢性炎症以及很多癌症
cyclins 细胞周期蛋白
carcinoma [ˌkɑːsɪˈnəʊmə] n. 癌
cell cycle [sel ˈsaɪkl] 细胞周期。指细胞从一次分裂完成开始到下一次分裂结束所经历的全过程，分为间期与分裂期两个阶段
angiogenesis [ˌændʒɪəʊˈdʒenɪsɪs] 血管生成。指从已有的毛细血管或毛细血管后静脉发展而形成新的血管
endothelial [ˌɛndoʊˈθiːliəl] adj. 内皮的 n. 内皮
endometrial [ˌɛndoʊˌmɛtriəl] adj. 子宫内膜的
eukaryotic [uːkeəriˈɒtɪk] adj. 真核的，真核生物的
lymphoma [lɪmˈfəʊmə] n. 淋巴瘤
nanoparticle [ˈnænəʊpɑːtɪkl] 纳米粒子
metastasis [məˈtæstəsɪs] n. (癌细胞)转移。(癌细胞)转移意味着癌症扩散到与其最初所在不同的身体部位。转移最常发生于癌细胞脱离肿瘤并进入血流或淋巴系统时。癌细胞可以远离原始肿瘤并通过血流或淋巴系统在身体的不同部位沉淀和生长时形成新的肿瘤

matrix metalloproteinase（MMP）[ˈmeɪtrɪks][mɪtæləˈprəʊtiːnəs] 基质金属蛋白酶。因其需要 Ca^{2+}、Zn^{2+} 等金属离子作为辅助因子而得名的一个大家族。同一种 MMP 可降解多种细胞外基质成分，而某一种细胞外基质（ECM, extracellular matrix）成分又可被多种 MMP 降解，但不同酶的降解效率可不同。MMP 能降解 ECM 中的各种蛋白成分，破坏肿瘤细胞侵袭的组织学屏障，在肿瘤侵袭转移中起关键性作用

nasopharyngeal *adj.* 鼻咽的；咽之鼻部的

cancer stem cell（CSCs）[ˈkænsə(r) stem sɛl] 癌症干细胞。癌症干细胞是癌细胞的亚群，可以自我更新，在肿瘤块中产生多种细胞，并维持肿瘤发生。癌症研究人员认为肿瘤起源于癌症干细胞，这种细胞起源于正常干细胞的突变

recurrence [rɪˈkʌrəns] *n.* 复发

epithelial-mesenchymal transition（EMT）[ɛpɪˈθɛljəl͵mɛzɛˈkaɪməl trænˈzɪʃn] 上皮细胞-间充质细胞转换

pluripotency [pˈluərɪpətənsɪ] 多能性；万能分化性；全能性

gut microbiota [maɪkrəubaˈɪɒtə] 肠道菌群；肠道微生物

short-chain fatty acid 短链脂肪酸

colorectal cancer [͵koʊloʊˈrɛktəl ˈkænsər] 结直肠癌；大肠癌

adjuvant [ˈædʒʊvənt] therapy 辅助疗法

synergistic effect 协同效应

paclitaxel [͵pæklɪˈtæksəl] 紫杉醇，是一种有名的药物，具有广谱的抗癌性

herceptin 赫赛汀；单克隆抗体；贺癌平。是一种靶向 HER2 蛋白的单克隆抗体

doxorubicin [dɒksəʊˈruːbɪsɪn] *n.* 阿霉素（一种抗肿瘤药）

epidemiological studies [͵ɛpɪdiːmɪəˈlɒdʒɪk(ə)l ˈstʌdi] 流行病学研究。流行病学是研究特定人群中疾病、健康状况的分布及其决定因素，并研究防治疾病及促进健康的策略和措施的科学。是预防医学的一个重要组成部分，是预防医学的基础

Hazard ratio（HR）风险比。指某一种干预措施（试验组）的应用所产生的风险率与不用该干预措施或者空白对照以及安慰剂等对照时所产生的风险率的比值

Confidence interval（CI）置信区间。在统计样本量相同的情况下，置信水平越高，置信区间越宽

cohort study [ˈkəuhɔːt ˈstʌdi] 队列研究是将某一特定人群按是否暴露于某可疑因素或暴露程度分为不同的亚组，追踪观察两组或多组成员结局（如疾病）发生的情况，比较各组之间结局发生率的差异，从而判定这些因素与该结局之间有无因果关联及关联程度的一种观察性研究方法

meta-analysis [ˈmetəənˈæləsɪs] 荟萃分析是用统计的概念与方法，去收集、整理与分析之前学者专家针对某个主题所做的众多实证研究，希望能够找出该问题或所关切的变量之间的明确关系模式，可弥补传统的文献综述的不足

randomized controlled trial [ˈrændəmaɪzd kənˈtrəuld ˈtraɪəl] 随机对照试验是一种对医疗卫生服务中的某种疗法或药物的效果进行检测的手段，特别常用于医学、生物学、农学。随机对照试验的基本方法是，将研究对象随机分组，对不同组实施不同的干预，以对照效果的不同。具有能够最大程度地避免临床试验设计、实施中可能出现的各种偏倚，平衡混杂因素，提高统计学检验的有效性等诸多优点，被公认为是评价干预措施的"金标准"

思考题

1. 将下面的句子译成英文。

（1）不同的茶及其活性成分通过抑制增殖相关的信号转导途径来表现出对癌细胞选择性的毒性，进而预防癌症的形成。

（2）茶可以增强抗癌药物的功效，并且与其他化合物一起对几种癌症具有协同作用，这使得茶成为合适的佐剂。

（3）这些不一致的流行病学研究结果可能是不同的实验变量导致的，例如对茶摄入量不准确的评估、缺乏茶制作的标准、不同地区不同的茶叶成分和受试者之间的遗传差异。

2. 将下面的句子译成中文。

（1）Epicatechin-3-gallate (EGCG) could inhibit the proliferation and growth of bladder cancer cells via downregulating the expression of nuclear factor-kappa B (NF-κB) in both protein and mRNA levels.

（2）A study showed that green tea consumption could elevate levels of short-chain fatty acid-producing bacterial genera, and reduce bacterial lipopolysaccharide synthesis, and decrease the functional pathways abundance relevance to cancer, which may ameliorate the inflammation and inhibit colorectal cancer.

（3）In addition, the epidemiological studies have revealed that the tea consumption has inverse association with the risk of several cancers, though there are still some inconsistent results.

参考文献

[1] XU X Y, ZHAO C N, GAO S Y, et al. Effects and mechanisms of tea for the prevention and management of cancers: An updated review[J]. Achieves of Biochemistry and Biophysics, 2019. DOI: 10.1080/10408398.2019.1588223.

[2] XING L, ZHANG H, QI R, et al. Recent advances in the understanding of the health benefits and molecular mechanisms associated with green tea polyphenols[J]. Journal of Agricultural and Food Chemistry, 2019, 67: 1029-1043.

[3] YANG C S, ZHANG J. Studies on the prevention of cancer and cardiometabolic diseases by tea: Issues on mechanisms, effective doses, and toxicities[J]. Journal of Agricultural and Food Chemistry, 2019, 67: 5446-5456.

[4] LI X, YU C, GUO Y, et al. Association between tea consumption and risk of cancer: a prospective cohort study of 0.5 million Chinese adults[J]. European Journal of Epidemiology, 2019, 34: 753-763.

[5] SEN A, PAPADIMITRIOU N, LAGIOU P, et al. Coffee and tea consumption and risk of prostate cancer in the European Prospective Investigation into Cancer and Nutrition[J]. International Journal of Cancer, 2019, 144(2): 240-250.

Lesson 3   Effects on Metabolic Syndrome and Related Diseases

1. Reduction of body weight, alleviation of metabolic syndrome and prevention of diabetes

Overweight, obesity and type 2 diabetes are emerging major health issues in many countries. Metabolic syndrome (MetS) encompasses a complex of symptoms that include enlarged waist circumference and two or more of the following: elevated serum triglyceride, dysglycemia, high blood pressure and reduced high-density lipoprotein-associated cholesterol. Studies on possible beneficial effects of tea consumption on body weight reduction and MetS alleviation are summarized below.

1) Studies in animal models

The effects of tea and tea constituents on body weight and MetS have been studied extensively in animal models. Most of the studies showed that oral administration of green tea extracts (GTE) or (−)-epigallocatechin gallate (EGCG) significantly reduced the gaining of body weight or adipose tissue weight, lowered blood glucose or insulin levels, and increased insulin sensitivity.

Diet-induced liver steatosis is becoming a common disease, and its possible prevention by tea consumption warrants more investigation. EGCG treatment markedly reduces the severity of hepatic steatosis, liver triglycerides and plasma alanine aminotransferase concentration in mice fed with a high-fat diet.

2) Studies in humans

Systematic reviews and meta-analysis covering more than 26 earlier short-term randomized controlled trials (RCTs). Most of these studies used green tea or green tea extracts with caffeine, in studies for 8 to 12 weeks, on normal weight or overweight subjects. Some RCTs also showed that daily intake of Polyphenon E capsules (tea catechin preparation containing 400 or 800 mg EGCG), for 2 months by postmenopausal American women, decreased blood levels of LDL-cholesterol, glucose and insulin.

In contrast to the above described beneficial effects, two recent studies in British adults did not show such beneficial effects. The first study used a rather low dose of EGCG (200 mg daily). The second study used daily supplementation with green tea (>560 mg EGCG) plus caffeine (0.28 – 0.45 mg) for 12 weeks. In an RCT with 237 overweight and obese postmenopausal women in the United States, daily supplementation of 1,315 mg of GTE (843 mg EGCG) for 12 months had no effect on body weight, BMI or waist circumference; but decreased fasting insulin levels in those with elevated baseline levels. Reasons for these negative results are unknown and remain for further investigation.

3) Mechanistic considerations

There are many proposed mechanisms for the above described actions of tea polyphenols. They can be summarized into two major types of actions: ①the action of tea polyphenols in the gastrointestinal tract, and ②the systemic action of tea polyphenols in different organs. The combined effects would reduce body weight, alleviate MetS and reduce the risk of diabetes and cardiovascular diseases (CVDs).

Ingestion of green tea polyphenols has been shown to increase fecal lipid and total nitrogen contents, suggesting that polyphenols can decrease digestion and absorption of lipids and proteins.

The possibility that tea may affect gut microbiome has been studied in animal models and humans. For example, green tea powder feeding affected gut microbiota and reduced the levels of body fat, hepatic triglyceride and hepatic cholesterol; the reduction was correlated with the amount of *Akkermansia* and/or the total amount of bacteria in the small intestine.

In humans, green tea consumption has been reported to decrease the abundance of *Clostridium* species and increase the abundance of the *Bifidobacterium* species in fecal samples. Increase of intestinal *Bifidobacterium* by a prebiotic (*oligofructose*) has been shown previously to decrease biomarkers for diabetes in mice. These results suggest the possibility that tea may alleviate MetS by enriching the probiotic population in the intestine.

Many studies showed that ingestion of tea catechins suppressed gluconeogenesis and lipogenesis and enhanced lipolysis in a coordinated manner. These results suggest that the actions of tea catechins are mediated by energy sensing molecules, possibly AMP-activated protein kinase (AMPK). In response to falling energy status, AMPK is activated to inhibit energy consuming processes and promote catabolism to produce ATP. The activation of AMPK by EGCG and green tea polyphenols has been demonstrated *in vivo* and *in vitro*. There are also reports indicating that AMPK was activated in adipose tissues and skeletal muscle by green tea, black tea, oolong tea and Pu-erh teas.

2. Lowering of blood cholesterol, blood pressure and incidence of cardiovascular diseases

1) Studies in humans

The alleviation of MetS by tea is expected to reduce the risk for CVDs. Correlation between consumption of tea and decreased risk of stroke were reported by two studies from China and Japan. A meta-analysis of 14 prospective studies, covering 513,804 participants with a median follow-up of 11.5 years, found an inverse association between tea consumption and risk of stroke, and the protective effect of green tea appeared to be stronger than that of black tea. Many, but not all, studies in the U.S. and Europe demonstrated an inverse association between black tea consumption and CVDs risk.

Green tea has been shown to decrease plasma cholesterol levels and blood pressure as well as improve insulin sensitivity and endothelial function in humans. A systematic review and meta-analysis of 10 trials (834 participants) on the effects of green tea on blood pressure in pre-hypertensive and hypertensive individuals showed significant reductions in systolic and diastolic blood pressure with tea consumption.

In a multiethnic study of men and women (6,508 participants) in the U.S., over a median follow-up of 11.1 years for cardiovascular events, regular tea drinking was associated with a lower coronary artery calcification and lower incidence of cardiovascular events, while coffee drinking was associated with an increased incidence of cardiovascular events, compared with non-drinkers. Similarly, in the Dongfeng-Tongji cohort study in China involving 19,471 participants followed for 3.3-5.1 years, green tea consumption was associated with a reduced risk of coronary heart disease incidence and related biomarkers in middle-aged and older Chinese population.

2) Possible mechanisms

Beneficial effects of tea catechins in lowering plasma cholesterol levels, preventing hypertension and improving endothelial function contribute to the prevention of CVDs. The cholesterol lowering effect is likely due to the decrease of cholesterol absorption or reabsorption by catechins as well as the decrease of cholesterol synthesis via the inhibition of 3-hydroxy-3-methylglutaryl-CoA reductase (mediated by the activation of AMPK). Enhanced nitric oxide signaling has been suggested as a common mechanism for catechins to decrease blood pressure and the severity of myocardial infarction. Several studies have shown that green tea or black tea polyphenols increased endothelial nitric oxide synthase activity in bovine aortic endothelial cells and rat aortic rings.

3. Conclusion

Many laboratory studies strongly suggest the beneficial health effects of tea consumption in the prevention of chronic diseases. However, the results from epidemiological studies have not been consistent. Large cohort studies have shown beneficial effects in reducing the risk of deaths from all causes (combined) and CVDs, but that from cancer is inconsistent, and the protective effects are generally strong among nonsmokers. Concerning human intervention studies, there are many successful examples that demonstrated the beneficial effects of tea constituents in weight reduction and MetS alleviation, even though there are still some inconsistencies. On the other hand, intervention trials for cancer prevention have not yielded convincing results, possibly due to the difficulties in conducting long-term studies in the appropriate populations.

词汇表

metabolic syndrome (MetS) [ˌmetəˈbɒlɪk ˈsɪndrəʊm] 代谢综合征是指人体的蛋白质、脂肪、碳水化合物等物质发生代谢紊乱的病理状态，是一组复杂的代谢紊乱症候群，是导致糖尿病心脑血管疾病的危险因素。其具有以下特点：①多种代谢紊乱集于一身，包括肥胖、高血糖、高血压、血脂异常、高血黏、高尿酸、高脂肪肝发生率和高胰岛素血症，这些代谢紊乱是心、脑血管病变以及糖尿病的病理基础。可见糖尿病不是一个孤立的病，而是代谢综合征的组成部分之一；②有共同的病理基础，目前多认为它们的共同原因就是肥胖尤其是中心性肥胖所造成的胰岛素抵抗和高胰岛素血症；③可造成多种疾病增加，如高血压、冠心病、脑卒中甚至某些癌症，包括与性激素有关的乳腺癌、子宫内膜癌、前列腺癌，以及消化系统的胰腺癌、肝胆癌、结肠癌等；④有共同的预防及治疗措施，防治住一种代谢紊乱，也就有利于其他代谢紊乱的防治

diabetes [ˌdaɪəˈbiːtiːz] n. 糖尿病。指一组以高血糖为特征的代谢性疾病。高血糖则是由于胰岛素分泌缺陷或其生物作用受损，或两者兼有引起。糖尿病时长期存在的高血糖，导致各种组织，特别是眼、肾、心脏、血管、神经的慢性损害、功能障碍

waist circumference [weɪst səˈkʌmfərəns] 腰围

triglyceride [traɪˈɡlɪsəˌraɪd] n. 甘油三酯

dysglycemia [diːzɡlɪˈsiːmjə] n. 血糖代谢紊乱

high-density lipoprotein 高密度脂蛋白

cholesterol [kəˈlestərɒl] n. 胆固醇

warrants more investigation 值得进一步调查

steatosis [ˌstiːəˈtəʊsɪs] n. 脂肪变性。是指在变性细胞的细胞浆内，出现大小不等的游离脂肪小滴

aminotransferase [æmaɪnəˈtrænsfˈreɪz] n. 转氨酶

gastrointestinal [ˌɡæstrəʊɪnˈtestɪnl] adj. 胃肠的

Akkermansia Akk 菌，是人体肠道中一种可降解黏蛋白的益生细菌，它与肥胖、糖尿病、炎症和代谢紊乱呈负相关；它发挥益生作用的机制可能是调节肠道内黏液厚度和维持肠道屏障的完整性

clostridium 梭状芽孢杆菌

bifidobacterium 双歧杆菌

prebiotic 益生元

probiotic 益生菌

catabolism [kəˈtæbəlɪzəm] n. 分解代谢

myocardial infarction [ˌmaɪəʊˈkɑːdɪəl ɪnˈfɑːkʃn] 心肌梗死

思考题

1. 将下面的句子译成英文。

（1）摄入绿茶多酚已被证明可以增加粪便脂质和总氮含量，表明多酚可以减少脂质

和蛋白质的消化和吸收。

（2）关于人体干预研究，尽管仍有一些不一致之处，但有许多成功的例子证明了茶叶成分在减轻体重和减轻代谢综合征方面的有益作用。

（3）大多数研究表明，口服绿茶提取物或表没食子儿茶素没食子酸酯可显著降低体重或延缓脂肪组织重量的增加，降低血糖或胰岛素水平，并提高胰岛素敏感性。

2. 将下面的句子译成中文。

（1）Metabolic syndrome（MetS）encompasses a complex of symptoms that include enlarged waist circumference and two or more of the following: elevated serum triglyceride, dysglycemia, high blood pressure and reduced high-density lipoprotein-associated cholesterol.

（2）Beneficial effects of tea catechins in lowering plasma cholesterol levels, preventing hypertension and improving endothelial function contribute to the prevention of cardiovascular diseases.

（3）Many laboratory studies strongly suggest the beneficial health effects of tea consumption in the prevention of chronic diseases. However, the results from epidemiological studies have not been consistent.

参考文献

[1] YANG C S, WANG H, SHERIDAN Z P. Studies on prevention of obesity, metabolic syndrome, diabetes, cardiovascular diseases and cancer by tea[J]. Journal of Food and Drug Analysis, 2018, 26(1):1-13.

[2] LI X, WANG W, HOU L, et al. Does tea extract supplementation benefit metabolic syndrome and obesity? A systematic review and meta-analysis[J]. Clinical Nutrition, 2019. DOI: 10.1016/j.clnu.2019.05.019.

[3] YI M, WU X, ZHUANG W, et al. Tea consumption and health outcomes: umbrella review of meta—analyses of observational studies in humans[J]. Molecular Nutrition & Food Research, 2019. DOI: 10.1002/mnfr.201900389.

[4] YUAN F, DONG H, FANG K, et al. Effects of green tea on lipid metabolism in overweight or obese people A meta-analysis of randomized controlled trials[J]. Molecular Nutrition & Food Research, 2017, 62(1): 1601122.

[5] LI D, WANG R, HUANG J, et al. Effects and mechanisms of tea regulating blood pressure: evidence and promises[J]. Nutrients, 2019, 11: 1115.

Lesson 4 Effects on Aging

As the number of elderly rapidly increases, the number of patients with dementia is also increasing. The most important risk factor for dementia is "aging" while "senescence" acts as a promoting factor. Previous research on tea constituents caffeine, L – theanine, and epigallocatechin gallate (EGCG) repeatedly demonstrated benefits on mood and cognitive performance. These effects were observed when these phytochemicals were consumed separately and in combination.

Caffeine was found to mainly improve performance on demanding long-duration cognitive tasks and self-reported alertness, arousal, and vigor. Significant effects already occurred at low doses of 40 mg. L-theanine alone improved self-reported relaxation, tension, and calmness starting at 200 mg. L-theanine and caffeine combined were found to particularly improve performance in attention-switching tasks and alertness, but to a lesser extent than caffeine alone. Moreover, L-theanine was found to lead to relaxation by reducing caffeine induced arousal.

1. Catechins against aging and dementia

Oxidative stress has been related to aging and age related disorders. It is found that in a wide variety of pathological processes, including cancer, atherosclerosis, neurological degeneration, Alzheimer's disease (AD), ageing and autoimmune disorders, oxidative stress has its implications. Catechins have powerful antioxidative activity and been reported to be useful in combating aging and age related disorders like cancer, cardiovascular disorders and neurodegenerative diseases.

Oxidative damage, brain atrophy, and cognitive decline are suppressed in aged mice that ingest green tea catechins, and when ingested from a middle age, age-related cognitive decline is significantly suppressed. In addition, many studies of green tea catechins using experimental animals suggest that they can protect the brain from AD. Daily consumption of several cups of green tea is thus expected to be effective in the prevention and reduction of brain senescence and dementia.

AD is a brain disease associated with aging. Nerve cells die in AD, making it difficult for the patient to make correct decisions or retain memories. In humans, several studies showed that the consumption of green tea was associated with a low prevalence of cognitive impairment. In Japanese residents older than 60 years of age, a population-based prospective study was carried out. During a follow-up period of about 5 years, the multiple-adjusted odds ratio for the incidence of dementia was significantly lower in individuals who had consumed green tea daily compared with those who consumed green tea for several or no days in a week. No association was observed between the consumption of coffee or black tea and the incidence of dementia. Green tea extract may modulate brain activity in the prefrontal cortex, a key area that mediates working memory processing in the human brain. To prevent senescence and dementia, further studies into the mechanism of green tea catechins are needed in the future.

2. Green tea against neurodegeneration

Tea has been associated with many mental benefits, such as attention enhancement, clarity of mind, and relaxation. It has been demonstrated that green tea polyphenols prevent age-related neurodegeneration through improvement of endogenous antioxidant defense mechanisms, modulation of neural growth factors, attenuation of neuroinflammatory pathway, and regulation of apoptosis (Figure 8.4).

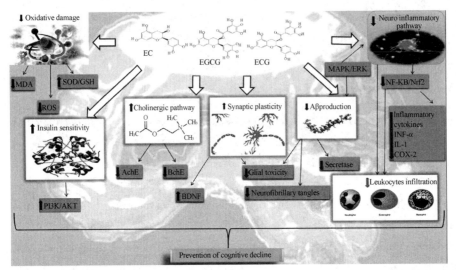

Figure 8.4 Neuropharmacological mechanisms of green tea catechins in prevention of cognitive decline

扫码看彩图

Moreover, the daily consumption of low concentrations of caffeine is beneficial to mood and mental performance, and reduces damage due to epilepsy. Caffeine is also known to improve the neurotransmission efficiency of dopaminergic neurons. Theanine, a tea component, reduces cell death due to various factors and helps prevent cranial nerve cell death. Administration of theanine caused a significant increase in dopamine concentrations within the brain, especially in the striatum, hypothalamus and hippocampus.

3. Effects of high-molecular-weight polyphenol (mitochondria activation factor) derived from black tea and oolong tea on mitochondria function

Mitochondria activation factor (MAF) is a high-molecular-weight polyphenol purified from oolong and black teas that increases mitochondrial membrane potential. MAF increases aerobic metabolic capacity in murine skeletal muscle, activates exercise training-induced intracellular signaling pathways that involve AMP-activated protein kinase (AMPK) and glucose transporter-4 (GLUT4), and improves endurance capacity. MAF also increases swimming velocities in sea urchin sperm by up to 8%, to the same extent as sperm-activating peptides secreted by the egg. These findings suggest that MAF is associated with the activation of glycolysis and lipolysis, which provide fuels for energy metabolism in mitochondria. The anti-hyperglycemic and anti-hyperlipidemic effects of MAF were evaluated in severe type 2 diabetic (db/db) mice. The results indicated that oral administration of MAF for 10 weeks reduced increases in fasting blood glucose levels, and it also reduced the accumulation of hepatic lipids and plasma lipids, even though these mice had greater food intakes than the control mice. In addition, MAF was found to be more effective than epigallocatechin gallate. Thus, MAF derived from oolong and black teas promotes glycolysis and lipolysis in hepatocytes and dramatically improves fatty liver.

4. Effects of theanine and other ingredients of tea on stress and aging

Stress is one of the most potent environmental factors known to accelerate brain aging. Suppression or prevention of stress-induced alterations is a potential therapeutic strategy for healthy aging. To clarify the mechanism of stress-induced brain aging, a new experimental animal model of psychosocial stress using male animal's territoriality has been developed. Theanine intake suppressed stress-derived adrenal hypertrophy, which is a sensitive stress-responsive phenomenon. In chronically stressed aged mice, theanine intake suppressed stress-induced disadvantages such as shortened lifespan, cerebral atrophy, cognitive dysfunction, and depression. The anti-stress effect of theanine was observed not only in mice but also in humans. However, the anti-stress effect of theanine was blocked by two other main components of tea, caffeine and epigallocatechin gallate (EGCG). On the other hand, epigallocatechin (EGC) and arginine (Arg) cooperatively abolished the counter-effect of caffeine and EGCG on psychosocial stress in mice. These results suggest that drinking green tea exhibits anti-stress effects when the contents of theanine, EGC, and Arg are relatively high and those of caffeine and EGCG are low.

5. Tea and bone health

With the increased aging population worldwide, strategies for prevention rather than treatment of chronic disease, such as osteoporosis, are essential for preserving quality of life and reducing health care costs. There is strong evidence from human studies that habitual tea consumption is positively associated with higher bone mineral density (BMD) at multiple skeletal sites, while the association with fracture risk is less clear. Fracture studies demonstrate a reduction or no difference in fragility fracture with tea consumption. There are key questions that need to be answered in future studies to clarify if higher consumption of tea not only supports a healthy BMD, but also reduces the risk of fragility fracture. And if the latter relationship is shown to exist, studies to elucidate mechanisms can be designed and executed.

词汇表

aging *n.* 变老；老化 *adj.* 变老的；老化的 *v.* 变老；使变老；（使）成熟，变陈
dementia [dɪˈmenʃə] *n.* 痴呆
senescence [sɪˈnesns] *n.* 衰老；老年化
mood and cognitive performance 情绪和认知表现
alertness, arousal and vigor 警觉、唤醒和活力
relaxation, tension and calmness 放松、紧张和平静
age related disorders 年龄相关疾病
Alzheimer's disease (AD) 阿尔茨海默病。临床上以记忆障碍、失语、失用、失认、视空间技能损害、执行功能障碍以及人格和行为改变等全面性痴呆表现为特征，病因迄今未明
neurodegenerative disease 神经变性疾病；神经退行性疾病。神经退行性疾病发病所伴随的病理变化是不可逆的，在患者出现认知障碍时，病程往往已到中晚期，此时治疗只能减缓疾病的发展，不能从根本上逆转神经网络的损伤
brain atrophy [breɪn ˈætrəfi] 脑萎缩
prevalence [ˈprevələns] *n.* 流行
prefrontal cortex [priˈfrʌntəl ˈkɔːrteks] 额叶前皮质；大脑前额叶皮层
epilepsy [ˈepɪlepsi] *n.* 癫痫
cranial nerve [ˈkreɪniəl nɜːv] 脑神经；颅神经
dopamine [ˈdəʊpəmiːn] *n.* 多巴胺。多巴胺是大脑中含量最丰富的儿茶酚胺类神经递质，调控中枢神经系统的多种生理功能。多巴胺系统调节障碍涉及帕金森病，精神分裂症注意力缺陷多动综合征和垂体肿瘤的发生等。多巴胺这种脑内分泌物和人的情欲、感觉有关，它传递兴奋及开心的信息。另外，多巴胺也与各种上瘾行为有关
striatum [strɪˈeɪtəm] 纹状体
hypothalamus [ˌhaɪpəˈθæləməs] 下丘脑
hippocampus [ˌhɪpəˈkæmpəs] 海马体
mitochondria activation factor (MAF) 线粒体激活因子
sea urchin [ˈsiː ɜːtʃɪn] *n.* 海胆
adrenal hypertrophy [əˈdriːnl haɪˈpɜːtrəfi] 肾上腺肥大
counter-effect 反作用
osteoporosis [ˌɒstiəʊpəˈrəʊsɪs] *n.* 骨质疏松；骨质疏松症

思考题

1. 将下面的句子译成英文。

（1）阿尔茨海默病和帕金森病等神经退行性疾病在现代人群中发病率很高，这是一类包括氧化应激、炎症、蛋白质集聚物积累等一系列毒性反应造成神经元死亡而引起的疾病。

（2）帕金森病（Parkinson's disease）是一种进行性中枢神经系统退化病，它是由人脑

多巴胺神经元分化而生的脑细胞的缺失而引起的，目前尚无法治愈。有研究显示，长期饮用绿茶和咖啡可以降低帕金森病的发病率。绿茶中的主要成分茶多酚和咖啡的主要成分咖啡因可以增加脑内多巴胺的含量、抑制神经毒素，从而对多巴胺能神经元具有保护作用，长期饮用的人不易患上帕金森病。

（3）茶叶中含有茶多酚、茶氨酸和咖啡碱等多种活性成分，在体内外试验中均表现出抗衰老作用，且茶叶能有效预防及改善心血管疾病、神经退行性疾病以及Ⅱ型糖尿病等与年龄相关的疾病。

2. 将下面的句子译成中文。

（1）Mitochondria activation factor (MAF) is a high-molecular-weight polyphenol purified from oolong and black teas that increases mitochondrial membrane potential.

（2）In chronically stressed aged mice, theanine intake suppressed stress-induced disadvantages such as shortened lifespan, cerebral atrophy, cognitive dysfunction and depression.

（3）There are key questions that need to be answered in future studies to clarify if higher consumption of tea not only supports a healthy BMD, but also reduces the risk of fragility fracture. And if the latter relationship is shown to exist, studies to elucidate mechanisms can be designed and executed.

参考文献

[1] HARA Y, YANG C S, ISEMURA M, et al. Health benefits of green tea-An evidence-based approach [M]. Boston, U.S: CABI, 2017: 178-184.

[2] DIETZ C, DEKKER M. Effect of green tea phytochemicals on mood and cognition [J]. Current Pharmaceutical Design, 2017, 23:2876-2905.

[3] FARZAEI M H, BAHRAMSOLTANI R, ABBASABADI Z, et al. Role of green tea catechins in prevention of age-related cognitive decline [J]. Journal of Cellular Physiology, 2019, 234:2447-2459.

[4] NASH L A, WARD W E. Tea and bone health: Findings from human studies, potential mechanisms, andidentification of knowledge gaps [J]. Critical Reviews in Food Science and Nutrition, 2017, 57(8):1603-1617.

Lesson 5 Bioavailability of Tea Components

1. Bioavailability of tea catechins

The potential bioactivity of green tea catechins *in vivo* is contingent upon their bioavailability and rates of biotransformation and elimination. The pharmacokinetics of green tea catechins have received considerable study resulting in the understanding that: ①catechin concentrations in the circulation peak at 1-5 μmol/L within 2-4 h following oral ingestion of tea or purified catechins; ②plasma half-lives of catechins are relatively short, ranging from 0.75 to 4.0 h following green tea ingestion; ③those catechins that are absorbed undergo extensive biotransformation through phase II xenobiotic metabolism resulting in numerous catechin metabolites that are glucuronidated, sulfated, and/or methylated; and ④ efflux transport proteins such as p-glycoprotein and multidrug resistance proteins (MRP)-1 and -2 limit enterocyte retention and/or apical-to-basolateral transport of catechins(Figure 8.5).

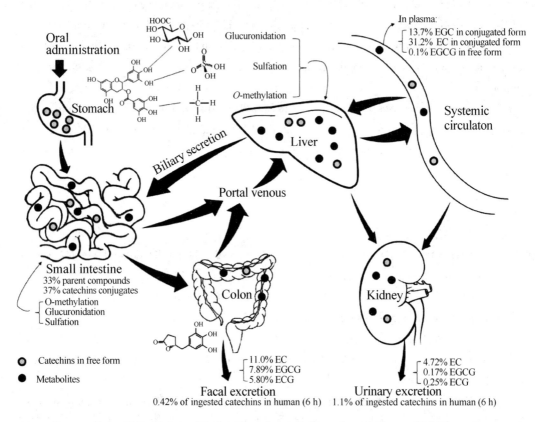

Figure 8.5 Schematic diagram of metabolism of green tea catechins

Further complicating the bioavailability of catechins is their susceptibility to auto-oxidation at near-neutral pH. This implicates not only low-acid foods/beverages, but also the human intestines as a potential site of conversion of monomers to dimerized products. Indeed, simulated gastric and intestinal digestion of EGCG, EGC and ECG results in substantial degradation of these compounds. For example, degradation of EGCG resulted in the formation of the auto-oxidation homodimers theasinensins A and D. Collectively, the oral bioavailability of green tea catechins is rather low resulting from a combination of instability during gastrointestinal digestion, relatively poor absorption, and rapid metabolism and elimination.

Interestingly, epidemiological evidence also suggests that tea consumption is correlated with reduced risk of gastric cancer.

Green tea is an extremely rich source of flavan-3-ol monomers. In a study, 500 mL of a bottled green tea was given as an acute supplement to ten volunteers after which plasma and urine were collected over a 24-hour period. The tea contained a total of 648 μmol of flavan-3-ols principally in the form of 257 μmol of EGC, 230 μmol of EGCG, 58 μmol of EC, 49 μmol of ECG and 36 μmol of (+)-GC.

Two of the native ECG and EGCG, were identified in plasma along with glucuronide, methyl-glucuronide and methyl-sulphate metabolites of (epi)gallocatechin and glucuronide, sulphate and methyl-sulphate metabolites of (epi)catechin. The C_{max} values ranged from 25 to 126 nmol/L and T_{max} values from 1.6 to 2.3 hours (Table 8.1). These T_{max} values and the pharmacokinetic profiles are indicative of absorption in the small intestine. The appearance of unmetabolised flavonoids in plasma is unusual. The passage of ECG and EGCG through the wall of the small intestine into the circulatory system without metabolism could be a consequence of the presence of the 3-O-galloyl moiety, as gallic acid per se is readily absorbed with a reported urinary excretion of 37% of intake.

Table 8.1 **Pharmacokinetic analysis of flavan-3-ols and their metabolites detected in plasma of ten volunteers following the ingestion of 500 mL of green tea**

Flavan-3-ol	C_{max}/(nmol/L)	T_{max}/h	$T_{1/2}$/h
(Epi)gallocatechin-O-glucuronide	126±19	2.2±0.2	1.6
4'-O-Methyl-(epi)gallocatechin-O-glucuronide	46±6.3	2.3±0.3	3.1
(Epi)catechin-O-glucuronide	29±4.7	1.7±0.2	1.6
(Epi)catechin-O-sulphates	89±15	1.6±0.2	1.9
4'-O-Methyl-(epi)gallocatechin-O-sulphates	79±12	2.2±0.2	2.2
O-Methyl-(epi)catechin-O-sulphates	90±15	1.7±0.2	1.5
(-)-Epigallocatechin-3-O-gallate	55±12	1.9±0.1	1.0
(-)-Epicatechin-3-O-gallate	25±3.0	1.6±0.2	1.5

Data expressed as mean value ± standard error ($n=10$).

Urine collected 0-24 hours after green tea ingestion contained an array of flavan-3-ol metabolites similar to that detected in plasma, except for the presence of minor amounts of three additional (epi)gallocatechin-O-sulphates and the absence of ECG and EGCG (Table 8.2). This indicates that the flavan-3-ols do not undergo extensive phase II metabolism. In total, 52.4 μmol of metabolites were excreted, which is equivalent to 8.1% of the ingested green tea flavan-3-ols. When the urinary (epi)gallocatechin and (epi)catchin metabolites are considered separately, a somewhat different picture emerges. The 33.3 μmol excretion of (epi) gallocatechin metabolites is 11.4% of the ingested EGC and (+)-GC, while the (19.1±2.2) μmol recovery of (epi)catechin represents 28.5% of intake (Table 8.2). These confirms that EC and (+)-C, in particular, are highly bioavailable being absorbed and excreted.

Table 8.2 Quantification of the major groups of flavan-3-ol metabolites excreted in urine 0-24 hours after the ingestion of 500 mL of green tea by ten volunteers

Flavan-3-ol metabolites (*number of isomers*)	0-24 h excretion/μmol
(Epi)gallocatechin-O-glucuronide (*1*)	6.5±1.2
4'-O-Methyl-(epi)gallocatechin-O-glucuronide (*1*)	4.4±1.5
4'-O-Methyl-(epi)gallocatechin-O-sulphates (*2*)	19.8±0.3
(Epi)gallocatechin-O-sulphates (*3*)	2.6±3.0
Total (*epi*)gallocatechin metabolites	33.3 (*11.4%*)
(Epi)catechin-O-glucuronide (*1*)	1.5±0.3
(Epi)catechin-O-sulphates (*2*)	6.7±0.7
O-Methyl-(epi)catechin-O-sulphates (*5*)	10.9±1.2
Total (*epi*)catechin metabolites	19.1 (*28.5%*)
Total flavan-3-ol metabolites	**52.4 (*8.1%*)**

Data expressed as mean value± standard error ($n = 10$). Italicised figures in parentheses indicate amount excreted as a percentage of intake.

The absence of detectable amounts of EGCG in urine, despite its presence in plasma, is difficult to explain. It is possible that the kidneys are unable to remove EGCG from the bloodstream, but if this is the case, there must be other mechanisms that result in its rapid decline after reaching C_{max}. Studies with rats have lead to speculation that EGCG may be removed from the bloodstream in the liver and returned to the small intestine in the bile. To what extent enterohepatic recirculation of EGCG, and also ECG occured in humans remain to be established.

Summing the C_{max} values for the individual plasma flavan-3-ols and metabolites in Table 8.1 results in an overall maximum plasma concentration of 538 nmol/L being attained after the ingestion of green tea. Despite the relatively low concentration of the green tea, flavan-3-ol metabolites in plasma (Table 8.1), the data on urinary excretion (Table 8.2) demonstrate that they are absorbed in substantial quantities, especially EC. Their failure to accumulate in comparable concentrations in plasma suggests that they are in a state of flux and are being more rapidly turned over in the circulatory system and, rather than accumulating, are excreted via the kidneys. In the circumstances, urinary excretion provides a more realistic assessment of absorption, but as this does not include the possibility of metabolites being sequestered in body tissues, this too is theoretically an underestimate of absorption, but to what degree remains to be determined. However, the fact that tissue sequestration has yet to be convincingly demonstrated suggests that it can only be at low levels, if at all.

A further point of note is that the plasma T_{max} times of the (epi)catechin metabolites following absorption in the small intestine are all in excess of 1.6 hours (Table 8.1), while with one exception, the T_{max} of the flavonol metabolites absorbed in the small intestine and derived from onion quercetin glucosides was 0.6–0.8 hour. This is unexpected, as in contrast to the onion flavonols, the green tea flavan-3-ols were already in solution and did not have to be solubilised post-ingestion. Furthermore, they were aglycones, not conjugates, and therefore, they did not have to be hydrolysed prior to absorption and metabolism. The delayed T_{max} of the green tea flavonols is unlikely to be due to a slower rate of absorption as their excretion is well in excess of the quantity of flavonol metabolites that appear in urine. Although further investigation is required, this does raise the possibility that ① the flavan-3-ols may be absorbed distal and flavonols in the proximal part of the small intestine or ② some component(s) in the green tea may be slowing transport through the gastrointestinal tract.

2. Bioavailability of L-theanine

After consumption of tea, L-theanine enters systemic circulation and is assumed to enter the brain. Several human studies indicate that L-theanine influences brain functioning. The absolute bioavailability of L-theanine, an amino acid, is expected to be close to 100%. Knowledge about the pharmacokinetics of L-theanine facilitates further study of this health effect. Total 15 healthy male volunteers with body weight (79.4±15.6) kg at 18-28 years old received 25-100 mg of L-theanine as tea, as L-theanine-enriched tea, and as biosynthetic L-theanine in aqueous solutions. Plasma was analysed for L-theanine content after which data were fitted with a 1-compartment model. For all interventions, the lag time was approximately 10 min and half-lives of absorption and elimination were approximately 15 and 65 min respectively. After approximately 50 min, maximum plasma concentrations of between 1.0 and 4.4 mg/L (M_W, 174.2; equivalent to 5.74 μmol/L and 25.27 μmol/L) were achieved. Maximum plasma concentration and area under the plasma-concentration-time curve (AUC) were dose-proportional (Table 8.3). This knowledge allows prediction of plasma concentrations for various dose regimens supporting further study of a health benefit of L-theanine.

3. Bioavailability of caffeine

The absolute bioavailability of orally administered caffeine was investigated in 10 healthy adult male volunteers, aged 18.8 to 30.0 years. The subjects were administered a 5 mg/kg dose of caffeine as either an aqueous oral solution or an intravenous infusion, on separate occasions about 1 week apart, in a randomized crossover fashion. Plasma samples were collected over the 24-h period following each dose and assayed for their caffeine content using a high-performance liquid chromatographic technique. The oral absorption was very rapid, reaching a peak (T_p) plasma concentration after (29.8±8.1) min. In addition, the variation in the maximum plasma concentration (C_{max}) was low, (10.0±1.0) μg/mL. The absolute bioavailability was assessed by comparing the areas under the plasma concentration vs. time curves for the intravenous and oral doses of caffeine. The rapid absorption resulted in essentially complete bioavailability of the oral caffeine, $F = (108.3±3.6)\%$. The caffeine plasma half-lives varied from 2.7 to 9.9 hours, indicating substantial inter-subject variability in its elimination.

Table 8.3 The pharmacokinetic profile of L-theanine.

L-Theanine Intervention	Parameter							
	Dose/ (mg)	$t_{1/2,a}$/ (min)	t_{max}/ (min)	$C_{max,70}$/ (mg/L)	$t_{1/2,e}$/ (min)	$AUC_{70}^{0\to\infty}$/ (min·g/L)	Cl/F_{abs}/ (L/min)	V_{hyp}/F_{abs}/ (L)
25 mg in 314 g H_2O_2	24.9±1.7	14±10	46±14	1.06±0.22	67±18	0.15±0.03	0.20±0.04	18±3
50 mg in 314 g H_2O_2	49.9±1.6	10±4	43±9	2.22±0.45	64±9	0.28±0.05	0.21±0.03	19±3
100 mg in 314 g H_2O_2	98.6±1.8	9±4	41±9	4.43±0.70	66±8	0.58±0.10	0.20±0.05	19±3
A cup of black tea of 314 g (naturally ~25 mg)	23.1±1.2	21±10	55±13	0.97±0.18	55±14	0.14±0.02	0.20±0.05	15±3
A cup of black tea of 314 g (naturally ~25 mg) + 25 mg pure compound	45.2±2.2	14±7	51±11	1.94±0.32	70±17	0.28±0.05	0.19±0.04	18±3

Values are means ± SD, $n = 15$; $t_{1/2,a}$, absorption half-life time; t_{max}, time of maximum plasma concentration; $C_{max,70}$, maximum plasma concentration corrected to a body weight of 70 kg; $t_{1/2,e}$, elimination half-life time; $AUC_{70}^{0\to\infty}$, area under the plasma concentration-time curve from $t = 0$ to ∞ min corrected to a body weight of 70 kg; Cl/F_{abs}, clearance over absolute bioavailability; V_{hyp}/F_{abs}, hypothetical distribution volume over absolute bioavailability.

Table 8.4 Oral absorption characteristics of caffeine in 10 young men

	Age/ (years)	Weight/ (kg)	Oral dose/ (mg/kg)	i.v. Dose/ (mg/kg)	Oral AUC_0^∞/ (min·μg/mL)	i.v. AUC_0^∞/ (min·μg/mL)	F/ (%)	T_p/ (min)	C_{max}/ (μg/mL)
Mean± SEM	21.8± 1.1	79.5± 3.9	4.94± 0.03	4.85± 0.09	4188.2± 554.9	4213.8± 547.1	108.3± 3.6	29.8± 8.1	10.0± 1.0
Range	18.8~ 30.0	64~ 106	4.84~ 5.17	4.64~ 5.63	1949.9~ 7424.9	2341.9~ 7440.5	93.8~ 129.7	3.6~ 79.8	6.9~ 16.2

AUC_0^∞ = the area under the plasma concentration vs. time curve from time zero to infinity [min·μg/mL]; F = the absolute bioavailability; T_p = the time at which the peak caffeine plasma concentration; C_{max} = maximum plasma concentration.

词汇表

contingent [kənˈtɪndʒənt] *adj.* 依情况而定的 *n.* 代表团；小分队

biotransformation *n.* 生物转化

pharmacokinetics [fɑːməkəʊkaɪˈnetɪks] 药物（代谢）动力学。简称药动学，主要研究机体对药物的处置（Disposition）的动态变化。包括药物在机体内的吸收、分布、生化转换（或称代谢）及排泄的过程，特别是血药浓度随时间变化的规律。药物的代谢与人的年龄、性别、个体差异和遗传因素等有关

half-life [ˈhɑːf laɪf] *n.* 半衰期

xenobiotic metabolism [ˌzenəʊbaɪˈɒtɪk məˈtæbəlɪzəm] 异生物质代谢

glucuronidated 葡糖醛酸化

sulfated [ˈsʌlfeɪtɪd] 硫酸化

methylated [ˈmeθɪleɪtɪd] 甲基化

efflux [ˈɛflʌks] transport protein 外排转运蛋白

p-glycoprotein and multidrug resistance protein *p*-糖蛋白和多药耐药蛋白

enterocyte [entərəˈsaɪt] *n.* 肠上皮细胞

apical-to-basolateral 顶端到基底外侧

biliary secretion [ˈbɪliəri sɪˈkriːʃn] 胆汁分泌

portal venous [ˈpɔːtl ˈviːnəs] 门静脉

systemic circulation [sɪˈstiːmɪk ˌsɜːkjəˈleɪʃn] 全身循环；体循环

fecal excretion [ˈfiːkl ɛksˈkriːʃən] 粪排泄

urinary excretion [ˈjʊərɪnəri ɛksˈkriːʃən] 尿排泄

intestine [ɪnˈtestɪn] *n.* 肠

simulated gastric and intestinal digestion 模拟胃肠消化

per se [ˌpɜː ˈseɪ] *adv.* 本身；本质上

enterohepatic recirculation 肠肝循环。指经胆汁或部分经胆汁排入肠道的药物，在肠道中又重新被吸收，经门静脉又返回肝脏的现象

sequestration [ˌsiːkwəˈstreɪʃn] 扣押；没收

gastrointestinal tract [ˌgæstrəʊɪnˈtestɪnl trækt] 胃肠道

1-compartment model 一室模型。指药物人体各组织器官的转运速率相同

area under the plasma-concentration-time curve 血浆浓度-时间曲线下面积。指单剂给药后，吸收进入血循环的药量可用 AUC 估算，AUC 单位为浓度×时间

dose-proportional 剂量比例

inter-subject variability 受试者间变异性

思考题

1. 将下面的句子译成英文。

（1）经口摄取的 EGCG 在肠道被吸收，通过门静脉进入肝。在此过程中，绝大多数

EGCG 会在肠黏膜或肝部发生共轭反应，同时也有一部分在肝部进一步被甲基化。接着，EGCG 代谢产物大多分泌至胆汁，也有一部分（包括完整的 EGCG）进入血液循环，在摄入 1~2 小时后达到峰值。

（2）咖啡因在进入人体后 45 分钟被胃和小肠完全吸收，一个健康成人的咖啡因的半衰期是 3~4 个小时。当某些个体患有严重的肝脏疾病时，咖啡因会累积，半衰期延长至 96 个小时。

（3）口服茶氨酸进入人体后通过肠道黏膜被吸收，进入血液，通过血液循环分散到组织器官，一部分在肾脏被分解后从尿中被排出。被吸收到血和肝脏的茶氨酸在 1 个小时后浓度下降，脑中的茶氨酸在 5 个小时后达到最高，24 个小时后人体中茶氨酸都消失了，以尿的形式排出。

2. 将下面的句子译成中文。

（1）Two of the native ECG and EGCG, were identified in plasma along with glucuronide, methyl-glucuronide and methyl-sulphate metabolites of (epi)gallocatechin and glucuronide, sulphate and methyl-sulphate metabolites of (epi)catechin. The C_{max} values ranged from 25 to 126 nmol/L and T_{max} values from 1.6 to 2.3 hours.

（2）Several human studies indicate that L-theanine influences brain functioning. The absolute bioavailability of L-theanine, an amino acid, is expected to be close to 100%.

（3）The rapid absorption resulted in essentially complete bioavailability of the oral caffeine, $F=(108.3\pm3.6)\%$. The caffeine plasma half-lives varied from 2.7 to 9.9 hours, indicating substantial inter-subject variability in its elimination.

参考文献

[1] BRUNO R S, BOMSER J A, FERRUZZI M G. Antioxidant capacity of green tea (*Camellia sinensis*)[M]// PREEDY V, Processing and impact on antioxidants in beverages. Oxford, U.K.: Academic Press, 2014; Chapter 4, 33-38.

[2] CLIFFORD M N, CROZIER A. Phytochemicals in teas and tisanes and their bioavailability[M]//Crozier A, Ashihar H, Tomas-Barberan F, Teas, cocoa and coffee-plant secondary metabolites and health. Chichester, U.K.: Wiley-blackwell, 2012; Chapter 3, 66-71;

[3] CAI Z Y, LI X M, LIANG J P. et al. Bioavailability of tea catechins and its improvement [J]. Molecules, 2018, 23:2346.

[4] PIJL P C, CHEN L, MULDER T P J. Human disposition of L-theanine in tea or aqueous solution [J]. Journal of Functional Foods, 2010, 2: 239-244.

[5] BLANCHARD J, SAWERS S J A. The absolute bioavailability of caffeine in man [J]. European Journal of Chinical Pharmacology, 1983, 24(1):93-98.

Lesson 6　Tea Safety

　　Although much of the available evidence regarding the beneficial effects of green tea has been provided by studies *in vitro* and in experimental model systems, controlled studies in humans are accumulating. In some cases, these clinical interventions are providing experimental evidence that directly support the profound health relations observed in epidemiological studies. More work is clearly needed to identify optimal intake levels and define safe strategies that optimize catechin bioavailability and bioefficacy.

　　The findings of epidemiologic studies suggest that green tea at ≥ 5–10 cups/day is associated with lower risks of liver-related morbidity and CVD-related mortality. However, several dozen anecdotal and case reports of idiosyncratic hepatotoxicity in humans have questioned the safety of green tea. Thus, it is possible that substantial inter-individual differences exist for catechin metabolism and/or that only supraphysiological doses are hepatotoxic. Moreover, because most case reports have been linked to dietary supplements, and dechallenge/rechallenge evidence is only provided in relatively fewer reports, product contamination remains a possibility to explain the acute hepatotoxicity of green tea. In contrast, intervention studies in otherwise healthy overweight men indicated that dietary supplementation of green tea polyphenols at 714 mg/day for 3 weeks did not affect numerous clinical chemistries for liver and kidney function. In addition, an analysis of several databases regarding adverse effects after consumption of dried GTE-containing supplements also concluded "**Overall, the average incidence (of adverse effects) combined over the last 10 years was approximately 0.00364998/10,000, which is considered extremely rare**".

　　A systematic review of the literature published up to 2013 concluded, based on an analysis of the reports from 34 randomized clinical trials, that liver-related adverse events were rare (seven cases in 1405 human subjects that had received dried GTEs versus one case in 1200 controls) and that liver-related adverse effects were generally "mild". Although it has put forward that a tolerable upper intake level of 300 mg EGCG/person/day is proposed by dietary supplements by Dekant W. *et al* (2018), it should be pointed out that this intake level of EGCG is not appropriate for green tea infusion and GTE-based beverages containing similar EGCG level with green tea infusion. In fact, there have been no reports of clinical toxicity when green tea is consumed as a beverage throughout the day. Tea, and therefore EGCG, is on the US FDA's list of compounds generally recognized as safe and approved for human consumption.

　　Polyphenon E, a decaffeinated supplemental preparation of tea catechins, contains approximately 65% EGCG and 30% other catechins. As tea catechins have poor bioavailability, most of the ingested EGCG does not reach the blood as a significant fraction is eliminated pre-systemically. In a phase I pharmacokinetic study of Polyphenon E, 20 human subjects were given single doses of either 200, 400, 600, or 800 mg EGCG, with each capsule containing 200 mg of EGCG and 68 mg of other catechins. Peak plasma EGCG levels of 200–400 ng/mL (0.4–0.8 μmol/L) could be achieved after the administration of these formulations at doses equivalent to the EGCG content of 8–16 cups of green tea.

A subsequent in-depth study of the safety and plasma kinetics of multiple-dose administration of purified EGCG and Polyphenon E was performed. This study examined once-daily and twice-daily dosing regimens of EGCG and Polyphenon E over a 4-week period in 40 healthy volunteers. In this study, standardized, defined, and decaffeinated green tea polyphenol oral products in amounts similar to the EGCG content in 16 Japanese-style cups of green tea were consumed once daily or in divided doses twice daily (4 capsules/day) for 4 weeks. EGCG intake at doses of 400 mg twice daily and 800 mg once daily established peak serum concentrations averaging 150–290 ng/mL. Peak concentrations were reached between 2.4 and 4.2 h.

Once-daily dosing of 800 mg resulted in approximately a 60% increase in the systemic exposure of free EGCG after chronic Polyphenon E administration. Purified EGCG and Polyphenon E were administered with food in these studies. All AEs (adverse events) during the 4-week period rated as mild and overall were similar to placebo. Abdominal discomfort (19%) was the most frequent AE. Other side effects included headache (6%), excess gas (6%) and dizziness (6%).

On the basis of the reported AEs and clinical laboratory data in this trial, the study agents and dosing schedules have been found to be safe and well tolerated by the study subjects for at least 1 month. The reported AEs were rated as mild events. The more common events include headache, stomach ache, abdominal pain, and nausea, which have been reported in subjects receiving green tea polyphenol treatment, as well as in subjects receiving placebo. There were no significant changes in blood counts and blood chemistry profiles after 4 weeks of green tea polyphenol treatment.

词汇表

morbidity and mortality [mɔːˈbɪdɪti ənd mɔːˈtæləti] 发病率与死亡率
anecdotal [ˌænɪkˈdəʊtl] adj. 轶事的；传闻的
idiosyncratic hepatotoxicity [ˌɪdɪəsɪŋˈkrætɪk ˌhepətoʊˌtɒkˈsɪsɪtɪ] 特异性肝毒性
inter-individual difference 个体间差异
dechallenge/rechallenge 去激发/再激发
polyphenon E 一种去咖啡因绿茶儿茶素混合物
dosing regimens [ˈdəʊsɪŋ ˈredʒɪmenz] 给药方案
adverse events (AEs) 副反应；不良反应

思考题

1. 将下面的句子译成英文。

（1）欧洲食品安全局（European Food Safety Agency）曾发布《绿茶儿茶素的安全性评价》报告，向大众发出健康警示：每日摄取 800mg 或以上的儿茶素膳食补充剂，可能会导致转氨酶（transaminase）升高，损伤肝脏。欧洲食品安全局也表示：根据安全评估，以传统方式冲泡的绿茶茶汤和绿茶饮料中的儿茶素含量一般被认为是安全的。

（2）对于健康的成年人，美国食品与药物管理局指出每天摄入 400mg 咖啡碱，也就是三到四泡茶叶是可以的，但是这却取决于个体对咖啡碱的敏感度以及个人身体的分解代谢速度。

（3）当你出现了失眠、抖动、焦虑、心率快、胃不舒服、恶心、头痛、烦躁不安等症状时，就说明你的咖啡碱摄入过多了，要注意减少。特别注意不要空腹饮茶。

2. 将下面的句子译成中文。

（1）Overall, the average incidence (of adverse effects) combined over the last 10 years was approximately 0.00364998/10,000, which is considered extremely rare.

（2）In fact, there have been no reports of clinical toxicity when green tea is consumed as a beverage throughout the day. Tea, and therefore EGCG, is on the US FDA's list of compounds generally recognized as safe and approved for human consumption.

（3）Abdominal discomfort (19%) was the most frequent AE. Other side effects included headache (6%), excess gas (6%) and dizziness (6%).

参考文献

[1] BRUNO R S, BOMSER J A, FERRUZZI M G. Antioxidant capacity of green tea (*Camellia sinensis*) [M]//PREEDY V. Processing and impact on antioxidants in beverages. Oxford, U.K.: Academic Press, 2014: 33-38.

[2] YATES A A, Jr. ERDMAN J W, SHAO A, et al. Bioactive nutrients-Time for tolerable upper intake levels to address safety [J]. Regulatory Toxicology Pharmacology. 2017, 84: 94-101.

[3] DEKANT W, FUJII K, SHIBATA E, et al. Safety assessment of green tea based beverages and dried green tea extracts as nutritional supplements [J]. Toxicology Letters, 2017, 277: 104-108.

[4] NANCE C L. Clinical efficacy trials with natural products and herbal medicines [M]//RAMZAN I. Phytotherapies Efficacy, Safety, and Regulation. New Jersey, U.S.: John Wiley & Sons, 2015: 73-76.

Lesson 1 World Tea Production and Consumption

Tea is a worldwide traditional beverage, about 160 countries and regions around the world to enjoy tea products. The tea market has been formed for more than a hundred years and has maintained stable growth as a whole. At the same time, it is also faced with a competitive buyer's market state in which the production exceeds the consumption. Therefore, it is an important basis for tea marketing and tea export and import trade activities to analysis the consumption situation of tea market at home and abroad, which is of guiding significance to the healthy and sustainable development of tea market.

At present, more than 60 countries in the world have grown and processed tea, more than 160 countries and regions have tea consumption habits, about hundreds of millions of people drink tea every day, the tea industry has attracted worldwide attention and has lasted for thousands of years, tea production and consumption complement and interact with each other.

1. World tea production

The global tea production has increased more than tenfold over the past century. The world tea production is mainly distributed in Asia and Africa, mainly in developing countries, accounting for nearly 97% of the world's total output. Asian tea–producing countries are mainly concentrated in China, Sri Lanka and India. Africa are mainly in East Africa, a few in Central and South Africa, with Kenya, Malawi, Uganda, Tanzania and Mozambique are all major tea producers, which account for 91% of Africa's total tea production.

1) Tea plantation area and yield

Since 2008, the world tea plantation and production have been increasing gradually. According to the latest statistics from the International Tea Committee (ITC), in 2017, the global tea planting area reached 4.89 million hectares and tea production was 5.81 million tons, an increase of 43.0% and 46.6% respectively over 2008 (Table 9.1). Asia and African remained the world's leading producers, accounting for 87% and 11.4% of the world's total output respectively.

In recent years, with the expansion of the tea planting area and the increase in the utilization rate of tea per unit area in the major tea-producing countries, the world tea production has maintained a steady growth with the planting area for ten years (Table 9.1), the world tea planting area has increased by 1.47 million hectares, up 43.0% from 3.42 million hectares to 4.89 million hectares. The annual growth rate ranged from 3.4% to 4.8%, and the average annual growth rate was 4.04%. The tea yield increased by 1.47 million tons from 3.965 million tons to 5.812 million tons, and the annual growth rate ranged from 1.4% to 6.8%, the average annual growth rate was 4.37%.

Table 9.1　　The area and yield of tea planted in the world from 2008 to 2017

Units: ten thousand hectares (area); 10,000 tons (yield)

Year	2008	2009	2010	2011	2012	2013	2014	2015	2016	2017
Area	342	355	368	384	399	418	437	452	472	489
Annual growth rate/%	3.8	3.7	4.3	3.9	4.8	4.5	3.4	4.4	3.6	—
Yield	396.5	401.9	428.1	457.1	469.3	500.2	520.9	528.5	557.4	581.2
Annual growth rate/%	1.4	6.7	6.8	2.6	6.6	4.2	1.5	5.3	4.3	—

In 2017, among the world tea plantation area of the top ten countries, China and India account for three-quarters of the world's total area (Table 9.2). Uganda's tea plantation area increased rapidly, surpassing Japan for the first time, and entering the world's top 10 plantation area, but its annual tea production was still lower than Japan (Table 9.3).

Table 9.2　　The world's top 10 tea plantation area countries in 2017

Countries	Plantation Area/ten thousand hectares	Countries	Plantation Area/ten thousand hectares
China	305.9	Indonesia	11.7
India	59.0	Myanmar	8.0
Kenya	23.3	Turkey	7.7
Sri Lanka	20.3	Bangladesh	5.9
Vietnam	13.4	Uganda	4.4

In 2017, world tea production reached 5.812 million tons, an increase of 46%, from 2008 to 2017. One of the biggest contributors was China (Table 9.3). China and India were still in the forefront of tea production, accounting for 67.6% of the world's total production, with the exception of Vietnam, Indonesia, Argentina and Bangladesh, the top 10 tea producing countries saw an increase from the previous year.

Table 9.3　　　　　The world's top 10 tea production countries in 2017

Countries	Production/10,000 tons	Countries	Production/10,000 tons
China	260.9	Vietnam	17.5
India	132.2	Indonesia	13.4
Kenya	44.0	Argentina	8.2
Sri Lanka	30.8	Bangladesh	7.8
Turkey	25.5	Japan	7.8

2) Types of tea production

The main types of tea produced in the world are black tea and green tea, from 2008 to 2017, their production is increasing year by year (Table 9.4). The growth rate of black tea and green tea were 35.5% and 73.4% respectively, the growth rate of green tea was greater than that of black tea. It shows that the consumption of green tea is on the rise from a worldwide. In 2018 German tea market, the proportion of green to black tea was 73% to 27%.

Table 9.4　　　　　World tea production in 2008 to 2017　　　Unit: 10,000 tons

Year	2008	2009	2010	2011	2012	2013	2014	2015	2016	2017
Green tea	116.4	123.7	126.9	137.1	148.7	156.0	166.6	173.7	182.5	201.8
Black tea*	280.1	278.2	301.2	320.0	320.6	344.2	354.3	354.8	374.9	379.4
Total	396.5	401.9	428.1	457.1	469.3	500.2	520.9	528.5	557.4	581.2

*: black tea contains black tea, white tea, yellow tea, and oolong tea.

3) Major tea producing countries

China with the largest variety of tea production mainly produces green tea, accounted for more than 60% of all tea types. Myanmar is mainly green tea and pickled tea, Japan mainly produces green tea, including steamed green tea (sencha), matcha, yulu tea (Gyokuro), etc; India and Kenya produce mainly CTC black tea and a small amount of green tea, while Vietnam produces green tea and scented tea. Sri Lanka and Turkey produce traditional black tea and a handful of green tea. Argentina produces broken black tea and Yerba Mate (made from the leaf of *Ilex paraguariensis*, not *Camellia sinensis*).

2. World tea consumption

Tea is a universal beverage. Tea drinking habits vary in different countries of world affected by geographic position, climate and dietary culture.

1) Current situation of tea consumption

In the past decade, the world tea consumption has maintained a growing trend. According to the statistics of the ITC, world tea consumption has increased by 1.736 million tons in the ten years from 2008 to 2017 (Table 9.5).

Table 9.5　　　　　World tea consumption from 2008 to 2017

Year	2008	2009	2010	2012	2013	2014	2015	2016	2017
Total/10,000 tons	383.5	390.9	415.4	453.8	470.2	487.9	503.5	528.3	557.1

Table 9.6 showed that China and India were the major tea-consuming countries with total tea consumption of 3.183 million tons, accounting for 57.1% of the world's total tea consumption. At the same time, the tea–producing countries still dominate the global tea consumption, the total consumption of China, India, Turkey, Japan and Indonesia were 3.63 million tons, accounting for 65.1% of the world. However, the tea consumption are going to keep on rising in tea consuming countries as Pakistan, Russia, the United States and Egypt's, which are very promising consuming markets. Turkey tea market is very difficult for other countries to enter because of its high tariffs of 145%. In the European Union market, the consumption shows a very small increase in the past decades, which mainly due to the Per capita consumption of tea in the UK declined from 1.9 kg in 2002 to 1.4 kg in 2016.

Table 9.6　　　　　The world's top 10 tea consumption countries in 2017

Countries	Total consumption/10,000 tons	Countries	Total consumption/10,000 tons
China	212.4	the United States	12.6
India	105.9	the United Kingdom	10.9
Turkey	25.0	Japan	10.4
Pakistan	17.5	Egypt	9.7
Russia	16.3	Indonesia	9.2

2) Tea consumption structure

Traditional tea consumption structure—World tea consumption mainly include black tea, green tea, oolong tea. The main tea consumption in the world is still black tea, accounting for about 80%, while green tea is only 14%. Morocco accounts for about 90% of green tea consumption, Japan's oolong tea about 29% and Germany's special tea about 10%.

World black tea consumption structure—Western Europe is the region with the largest and most common consumption of black tea, followed by Eastern Europe, the Asia–Pacific region, Africa and the Middle East. World green tea consumption structure: Northwest Africa is a traditional green tea consumption area, green tea consumption in Asia-Pacific region is very popular. Western Europe, North America green tea consumption still has a great potential.

Tea has become an inseparable part of Chinese life, traditional tea consumption is divided into five main types: green tea, scented tea, black tea, oolong tea and compressed tea. In 2018, green tea consumption remained dominant, accounting for 63.1% of the total market, black tea market (9.9%) continues to maintain, dark tea (14.0%) and white tea (1.5%) markets continue to expand, and oolong tea (11%) is still in a low period, traditional superior famous tea and reprocessed tea type increase.

Consumption structure of value-added tea—In recent years, with the rapid and healthy reprocessed tea has become the mainstream of consumption, the proportion of traditional bulk tea has declined, "tea bag" "instant tea" "tea drinks" and "healthy tea" have become more and more popular in the world. Tea bag, because of its characteristics of quantitative, safe, convenient and fast, which has accounted for more than 50% in Europe and the United States. Consumption of instant tea is the United States, the United Kingdom, Ireland and Japan, which occupies the first place in the United States. The shares of tea drinks and healthy tea in the market have increased, tea drinks with natural, healthy thirst quenching, refreshing characteristics, to meet the pursuit of a healthy lifestyle consumer trend, coupled with its elegant fragrance, rich health ingredients, which has a huge space for development.

3. Conclusion

In the last decade, global tea production and consumption have maintained steady increase, agricultural output has been increasing, but the contradiction between production and consumption is still prominent, and tea producers still dominate global tea consumption.

Although black tea is the main consumption of the world, in recent years, the proportion of green tea consumption with healthy awareness is keeping on the rise. With the change of consumption level and concept, the global consumption structure has become more diverse, the consumption structure has gradually shifted to middle and high-grade, convenient and fashionable. Especially, rising trend of urban population and consumer's concern for nutrition, health, quality of foods and environment has resulted a change in consumption of organic tea.

It is worth noting that pilot tea production have been launched in France, Germany, Italy, Holland, Portugal, Spain, Switzerland, the United Kingdom, the United States, Australia, New Zealand in recent years, showing an emerging interest in developing tea production in countries where such cultivation is not traditional.

词汇表

Malawi [mɑːˈlɑːwi] 马拉维(非洲国家)

Uganda [juːˈɡændə] 乌干达(非洲国家)

Tanzania [ˌtænzəˈniːə] 坦桑尼亚(非洲国家)

Mozambique [ˌməʊzæmˈbiːk] 莫桑比克(非洲国家)

International Tea Committee (ITC) 国际茶叶委员会

Vietnam [ˌvjetˈnæm] 越南

Bangladesh [ˌbaːŋɡləˈde] 孟加拉国

Myanmar [ˈmjænmɑː] 缅甸

matcha [ˈmætʃə] *n.* 抹茶, 即碾茶, 日本茶名

Yerba Mate [ˈjɛːbɚ] [meɪt] 马黛茶, 由马黛树的叶子制成

Pakistan [ˈpækistæn] *n.* 巴基斯坦(南亚国家)

tariff [ˈtærɪf] *n.* 关税; 收费表 *vt.* 定税率; 征收关税

special tea 特种茶, 在国际贸易中是指相对于红茶、绿茶之外的一些特色茶叶产品

Asia-Pacific region [ˈeɪʃə] [pəˈsɪfɪk] 亚太地区 亚洲及太平洋沿岸地区的简称

Value-added tea 增值茶类

bulk tea [bʌlk] 散茶

思考题

1. 将下面的句子译成英文。

(1) 目前, 全球已有60多个国家种植和加工茶叶, 160多个国家和地区的人们有茶叶消费习惯, 约30亿人每天饮用茶叶, 茶叶行业举世瞩目, 几千年来经久不衰。

(2) 茶是一种通用饮料, 世界茶叶消费一直保持着增长趋势, 由于受到国家、地理、气候和文化等许多因素的影响, 不同国家的饮茶习惯各不相同, 不同地区的人喜欢不同类型的茶。

(3) 2017年, 中国和印度是世界上最主要的茶叶消费国, 茶叶消费量为318.3万吨, 占世界茶叶消费量的57.1%。同时, 茶叶生产国在全球茶叶消费中仍占主导地位。

2. 将下面的句子译成中文。

(1) In recent years, with the expansion of the tea planting area and the increase in the utilization rate of tea per unit area in the major tea-producing countries, the world tea production has maintained a steady growth with the planting area for ten years.

(2) China was responsible for the accelerated growth in global tea output, as production in the country more than doubled from 1.17 million tons in 2007 to 2.44 million tons in 2016. The expansion in tea production in China was in response to unprecedented growth in domestic demand, underpinned by the country's economy which grew at an average annual rate of 10 percent over the last 30 years.

(3) At the world level, black tea production increased annually by 3.0 percent and green tea by 5.4 percent over the last decade, in response to continued firm prices and green tea's perceived health benefits.

参考文献

[1] China Tea Marketing Association. Development report on world tea production and marketing in 2017[J]. Tea World, 2018, 12: 24-36.

[2] China Tea Marketing Association. Analysis report on the situation of tea production and marketing in China in 2018[J]. Tea World, 2019(2): 10-15.

[3] DU X. Tea market and trade[M]. Beijing: China Culture and History Press, 2013: 21-23.

[4] Committee on commodity problems, Intergovernmental group on tea. CCP: TE 18/CRS1 Current market situation and medium term outlook[R]. Hangzhou, China: FAO, 2018.

Lesson 2 Tea Exports and Imports

1. Global tea imports and exports

1) Global tea imports

From 2008 to 2017, world tea import has remained relatively stable (Table 9.7). In 2017 the amount of tea imports was 1,718,000 tons, among which 489,600 tons were imported from Asian countries, 370,000 tons from African countries, and 137,700 tons from European countries, large importers were still Pakistan, Russia, the United States and Britain, followed by Turkey.

Table 9.7   Total world tea imports from 2008 to 2017

Year	2008	2009	2010	2011	2012	2013	2014	2015	2016	2017
Total/10,000 tons	157	150.3	165.6	167.2	166.5	173.5	174.1	172.2	173.5	171.8

Tea imports in the top 10 tea imported countries were fluctuated compared to the previous year (Table 9.8). Pakistan was the largest importer of tea, which mainly imported black tea. The second was Russia, which imported of 160,000 tons, in which 90% was black tea. However, the proportion of imported green tea and superior famous tea is increasing; the main suppliers were Kenya, India and Sri Lanka, taking up nearly 70% of Russia's total tea imports. The United States ranked third, mainly imported black tea from Argentina, India, Sri Lanka, Kenya, while the volume of green tea import proportion is increasing, most of which comes from China. In the tea-drinking countries of Europe, the UK is a non-tea producer and all tea consumed is imported, with the amount reaching 109,000 tons, in which Kenya was the main supplier accounted for 54.7% of the total.

Table 9.8　　　　　　　　The world's top 10 tea importers in 2017

Counties/Regions	Imports/10,000 tons	Year-on-year growth/%
Pakistan	17.5	0.7
Russia	16	0.0
the United States	12.6	-3.8
the UK	10.9	1.6
Commonwealth of Independent States(CIS)	8.8	1.5
Egypt	7.8	-13.4
Morocco	7.3	7.8
Iran	6.3	-4.3
Dubai	5.8	-6.9
Iraq	4.1	3.2

Most of the tea imported into the UK is bulk tea (mainly CTC tea and green tea), which comes to the market after being blended, packaged (packet tea) or processed into tea bag. Egypt imports bulk tea and CTC tea, mainly from Kenya, India and Sri Lanka. Morocco mainly imports high-grade eyebrow tea and gunpowder tea from China.

Dubai (the United Arab Emirates) is a tea distribution center in the Middle East and mainly imported black tea and tea bag from India and Sri Lanka. The teas imported in Iraq mainly include black tea and bulk teas from Sri Lanka.

2) Global tea exports

In 2017, the world tea exports were 1.791 million tons, decreased by 1.1% over 2016 (Table 9.9). Black tea exports were 1.403 million tons, accounting for 78.3%, and the green tea exports were 388,000 tons, accounting for 21.7%. The world exports of green tea increased by 6.7% compared with 362,000 tons in the previous year.

Table 9.9　　　　　　　　Total world tea exports from 2008 to 2017

Year	2008	2009	2010	2011	2012	2013	2014	2015	2016	2017
Total/10,000 tons	165.2	161.5	178.6	176.2	177.3	185.8	182.3	179.4	179.8	179.1

The exporting countries with a large decline in 2017 had Kenya, Uganda and Vietnam (Table 9.10). Chinese tea export value ranked first in the world followed by Sri Lanka and Kenya.

Kenya is the world's largest tea exporter (Table 9.10), with exports falling by 65,000 tons in 2017 reduced by 13.4% over 2016. Pakistan as the Kenya's largest tea importer received 153,000 tons, accounted for 37% of Kenya's total exports in 2017. China was the second largest tea exporter's exports with 8.1% growth in 2017. Sri Lanka's exports fell slightly to third place, mainly exported black tea to Asia and Russia. Other major exporters were: India, Vietnam, Argentina, India, Uganda, Malawi, Tanzania, etc.

Table 9.10   The world's top 10 tea exporters in 2017

Countries	Exports/10,000 tons	Year-on-year growth/%
Kenya	41.6	-13.4
China	35.5	8.1
Sri Lanka	27.8	1.0
India	24.7	10.2
Vietnam	14.0	-5.6
Argentina	7.49	2.1
Indonesia	5.42	6.9
Uganda	4.70	-10.3
Malawi	2.90	0.0
Tanzania	2.75	5.7
Global total exports	179.1	-1.1

2. Tea imports and exports in China

1) China's tea imports

With the rapid development of China's economy and the continuous improvement of residents' living standards, tea consumption has shown diversified and personalized trend, and tea imports have grown rapidly (Table 9.11). In 2017, China imported 29,800 tons of tea, worth of US$ 149.8 million, up 31.2% and 33.9% year-on-year respectively. Among them, black tea was still the main imported tea, about 25,000 tons, accounting for 85% of the total tea imports, the main suppliers were Sri Lanka and India. Secondly, the amount of imported green tea reached 2,152 tons, accounting for 8.9%. The main suppliers were Vietnam and India. Sri Lanka and India are the major two suppliers to China's tea imports, accounting for 55% in 2017 over 28% in 2010.

China's tea imports amounted to 1,900 tons and US$ 0.3 million in 1999, while 5,400 tons and US$ 1.8 million in 2008. In 2018, those increased to 35,500 tons and US$ 178 million. Black tea is still the main imported tea, accounting for 83.3%, followed by green tea and oolong tea, accounting for 8.9% and 6.5% respectively. According to the average import price, the overall average price is US$ 5.03/kg, while the average price of oolong tea and scented tea is higher, both exceeding US$ 13/kg.

Table 9.11　　　　　　　　China's total tea imports from 2013 to 2017

Year	2013	2014	2015	2016	2017
Total/10,000 tons	1.96	2.26	2.29	2.27	2.97

2) China's tea exports

China's tea exports had been on the rise (Table 9.12) during 2013-2017. In 2017, China exported 355,000 tons of tea, amounting to US$ 1.61 billion, with an average export price of US$ 4.54 per kilogram. China exported tea to 128 countries and regions, 12 of which exceeded 10,000 tons, accounting for 64.8% of the total annual export. In 2018, China's tea export volume was 365,000 tons and the export amount was US$ 1.78 billion, an increase of 10.5% over last year.

Table 9.12　　　　　　　　China's total tea exports in 2013-2017

Year	2013	2014	2015	2016	2017
Total/10,000 tons	32.58	30.15	32.49	32.87	35.53

From the point of export volume (Table 9.13), the five largest major provinces account for nearly 90% of the total, among which Zhejiang was occupied the largest export volume, accounts for nearly 50%, followed by Anhui province 17.9%. From the point of export value, Zhejiang accounts for 31.72%, Anhui 15.8%, Fujian 14.8%, Hubei 8.23%, Guangdong 6.0% and Hunan 5.58%.

Table 9.13　　　　　　　　The top 10 provinces of tea exports of China in 2017

Provinces	Exports/10,000 tons	Provinces	Exports/10,000 tons
Zhejiang	17.50	Jiangxi	1.061
Anhui	6.437	Yunnan	0.7925
Hunan	3.288	Henan	0.7282
Fujian	1.950	Guangdong	0.6749
Hubei	1.633	Chongqing	0.4868

China's export tea types mainly include green tea, black tea and special tea, etc. China's green tea exports are the largest in the world, accounted for more than 80% of the world's green tea trade. Black tea, oolong tea, scented tea and pu-erh tea, accounted for 10%, 4%, 2% and 1% respectively. China's exporting teas is still dominated by bulk loose tea. With the increase of production cost of labor force and raw materials, the export price of tea is raising year by year, which makes the competitiveness of China's tea export gradually declined.

In 2017, China exported tea to 128 countries and regions, with 12 countries more than 10,000 tons, accounting for 64.8% of the total annual exports. Currently, the traditional export markets of the green tea mainly include Morocco, Uzbekistan and Japan. Asia and Africa are still China's main target tea markets, the proportion of China's tea exports to Africa continued to rise, while that in Asia is on a downward trend.

3. Conclusion

Over the past decade, the global tea import and export have maintained a synchronous and steady growth, but the growth rate of import was slower than that of export.

At present, the export of tea products in China is mainly green tea, and the proportion of black tea exports has declined significantly. However, the share of special tea in the export of tea products in China still has tremendous room for growth. Therefore, it is necessary to optimize the structure of China's tea export. On the basis of ensuring and stabilizing the original traditional market, it should explore the international tea market with the continuous development of "one belt one road" initiative.

词汇表

the United Arab Emirates 阿联酋
Commonwealth of Independence States (CIS) 独联体
export value 出口额
packet tea 包装茶
export volume 出口量
synchronous ['sɪŋkrənəs] adj.同步的；同时的

思考题

1. 将下面的句子译成英文。

（1）巴基斯坦是最大茶叶进口国，茶叶进口以红茶为主，目前的主要茶叶供应国是肯尼亚，占比73.7%，绿茶主要从越南和中国进口。排名第二的国家为俄罗斯，进口量为16万吨，同比持平。

（2）英国进口的茶叶多数是散装茶，经过拼配、分装(小包装)或加工成袋泡茶之后进入市场。进口茶叶以CTC红碎茶为主，中国是英国主要的绿茶进口国。

（3）近十年来，全球茶叶进出口保持同步稳定增长，但进口增长率低于出口增长率。大多数产茶国的茶叶在国内消费，特别是中国和印度。世界茶叶出口量和茶叶总产量所占比例逐年下降。

2. 将下面的句子译成中文。

(1) In 2017, the world tea exports reached 1.791 million tons, which was slightly lower than the previous year's 1.798 million tons, among which the black tea export was 1.403 million tons, accounting for 78.3%.

(2) From the point of export tea type, China's export tea variety, mainly include green tea, black tea and special tea, among which green tea exports is the largest in the world, accounting for more than 80% of the world's green tea trade.

(3) Therefore, it is necessary to optimize the structure of China's tea export. On the basis of ensuring and stabilizing the original traditional market, it should explore the international tea market with the continuous development of "one belt one road" initiative.

参考文献

[1] China Tea Marketing Association. Analysis report on the situation of tea production and marketing in China in 2018[J]. Tea World, 2019(2): 10-15.

[2] LENG Y, SHANG H, SHI X, et al. A Brief Analysis of China's Tea Import and Export in 2018[J]. China Tea, 2019(4): 30-32.

[3] Committee on commodity problems, Intergovernmental group on tea. CCP: TE 10/CRS20 Market developments in selected countries—China[R]. New Delhi, India: FAO, 2010.

[4] Committee on commodity problems, Intergovernmental group on tea. CCP: TE 18/CRS1 Current market situation and medium term outlook[R]. Hangzhou, China: FAO, 2018.

Lesson 3 Tea Market Development and Outlook

1. Current development

There are more than 60 tea-producing countries and more than 195 tea-drinking countries in the world. In 2018, the total output of tea in the world was 5.856 million tons and consumption of tea reached 5.571 million tons. From 2008 to 2018, the growth rates were 52.7% and 46.6% respectively. China was responsible for the accelerated growth in global tea output, as production in the country more than doubled from 1.17 million tons in 2007 to 2.61 million tons in 2018. The expansion in tea production in China was in response to unprecedented growth in domestic demand, underpinned by the country's economy which grew at an average annual rate of 10 percent over the last 30 years. The massive expansion is also a result of increased consumer health consciousness and the rapid development of tea beverages in the country characterized by a longstanding tradition of drinking tea.

The import and export of world tea remains relatively stable, and black tea still occupies the main position in the world tea trade, accounting for more than 70%. In 2018, total tea export volume was 1.85 million tons, an increase of 60,000 tons with 3.43% growth over 2017; total global tea import volume was 1.73 million tons, an increase of 10,000 tons with 0.61% increase over 2017. African tea exports reached 655,000 tons, accounting for 35% of the total global tea exports. Especially, Kenya, Uganda and Malawi exported 475, 50 and 35 million tons, respectively.

At present, Pakistan is the largest tea importer in the world. In 2018, Pakistan imported 192,000 tons of tea, an increase of 9.58% over 2017. Pakistan mainly imports black tea, accounting for 99%. Kenya is the main tea supplier, importing 139,000 tons, accounting for 73%. Russia is the world's second largest tea importer, importing 153,000 tons of tea in 2018, a decrease of 6% over 2017. The United States imported 139,000 tons of tea, ranking third. Britain imported 126,000 tons of tea, ranking fourth. Kenya is Britain's largest tea supplier, exporting 74,000 tons. Generally, imports from major tea importing countries, such as Britain, the United States, Russia and Egypt, have slightly declined.

1) **Production continues to grow, the contradiction between production and consumption is prominent**

From 2008 to 2018, the world tea production has increased by 1.62 million tons, an increase of 52.7%. In 2018, statistics of the International Tea Commission showed that the global consumption of tea was 5.571 million tons, less than the global tea production of 5.856 million tons. The expansion of world tea consumption was underpinned by the rapid growth in per capita income levels, notably in China, India and other developing and emerging economies. Growth in demand expanded significantly in most of the tea producing countries in Asia, Africa and Latin America. In traditional importing countries of Europe (except Germany) and the Russian Federation, tea consumption has declined over the last decade. The European tea market is mature and per capita consumption has been declining as competition from other beverages, particularly bottled water and carbonated drinks, has intensified. This is obvious in UK (Figure 9.1).

This gap is also reflected in the tea trade. In 2017, global tea exports were 1.791 million tons, the growth rate of global tea exports was only 8.4% over the 10-year period. The ratio of world tea exports to tea production fell from 41.7% in 2008 to 30.8% in 2017. This further confirms the contradiction that the world tea production is greater than the consumption, and more measures are urgently needed to promote tea consumption.

2) **Trade barriers affect the development of the tea market**

In recent years, trade protectionism is becoming increasingly serious in the international market; On the other hand, due to the low price of tea exported by developing countries and the excessive supply in the export market, many importing countries have implemented the entry standards with the tendency of technical barriers against tea exporting countries, which greatly obstruct or restrict the entry of tea products. For example, the European Union, the

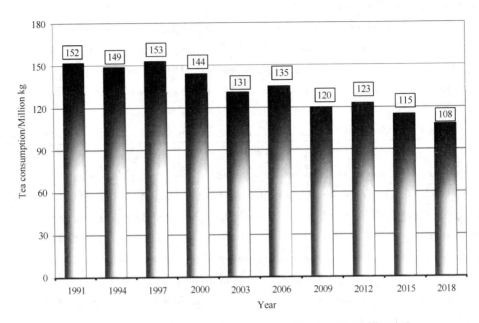

Figure 9.1　Annual apparent tea consumption in United Kingdom

United States, Japan, Morroco and other countries and regions have set green trade barriers to Chinese teas, including pesticide residues, system certification, packaging and label restrictions, which have a great impact on the export of Chinese teas.

2. Countermeasures and outlook

First, it shall moderately control the plantation and promote consumption. At present, while paying attention to the trend of global tea consumption, enterprises and industry organizations should strengthen the promotion of tea and health in various countries by holding various tea events to guide people to drink tea in a scientific and healthy way, and revitalize the global tea economy. At the same time, tea producing countries should moderately control the planting area of tea, improve the quality and control the yield, so as to reverse the sluggish growth of global tea trade.

Second, it is to establish the global tea sustainable development strategy. Global climate warming increases plant diseases and insect pests, in turn, the farmer's income reduces; the impact of maximum residue limits (MRL) of pesticide on tea exports, which affect the sustainable development of tea production and trade, urgently need to establish a sustainable development of global tea standard to protect tea garden environment to ensure tea safety. At the same time, the certifications of organic tea have different standards in different countries, which need to come to consensus on organic tea standard to maintain the healthy development of the tea trade.

Robust demand and associated high prices stimulated substantial supply increases over the past decade, resulting in significant growth in domestic consumption and trade. Black tea consumption by FAO is projected to grow at 2.5 percent annually to reach 4.17 million tons in 2027, reflecting the strong growth in consumption in producing countries, which should more than offset projected declines in traditional tea importing countries. The largest expansion within the five top producing countries is expected in China where an annual growth of 5.9% is projected over the next 10 years. African countries are expected to show higher growth in their consumption with Rwanda leading the move (9%) followed by Uganda (5%), Kenya (4.4%), Libya (4.4%), Morocco (4.2%) and Malawi (4.2%). Lower consumption growth rates are expected in western countries with UK consumption projected to be negative as black tea struggles to maintain consumers' interest amid growing competition from other drinks including coffee. Only Germany (1.4%) and Poland (1.3%), followed by the Netherlands and France (both at 0.6%) are expected to have consumption growth rates higher than the region's average of 0.2 percent.

Major factors contributing to the expansion in consumption in tea producing countries are the growth in per capita income, the increased awareness of the health benefits of tea consumption and the product diversification process attracting more customers in non-traditional segments including young people. The reactions to the rising per capita income in major emerging and developing countries and the growing awareness of tea health benefits were to stimulate consumption. Along with the improvement of people's health requirements, the pursuit of safety, health and quality teas, the area of organic tea gardens and ecological tea gardens will increase, the tea products will become diversified, and the comprehensive utilization of tea resources and the development of new products will be more adequate.

3. The world tea market to China's enlightenment

It is necessary to strengthen the quality and safety of tea export and improve the trading mechanism of tea market in China. In the aspect of exports, green trade barrier seriously hinders the development. It not only significantly reduces the export scale of Chinese tea and increases the export cost, but also seriously affects the product image of Chinese teas. Therefore, China should establish the assurance system of quality and safety, improve the proportion of organic tea gardens and ecological tea gardens, and sustainably develop physical control, biological control and other green prevention and control technologies to reduce the usage amount of pesticides and fertilizers.

Keeping up with the world tea consumption market dynamic, Chinese companies should enrich tea products to find new sources of growth. For the countries that traditionally consume black tea like Europe, America, some Asian countries, we should accelerate the production of novel black tea products to meet their demands.

With the increasing health consciousness, we should optimize the structure of export tea to promote our advantage on high-quality green tea, special teas including organic tea, oolong, white, yellow and dark teas. Moreover, the exchange and integration of different tea cultures will further promote the development of tea trade in different countries.

词汇表

emerging and developing countries 新兴和发展中国家

trade protectionism 贸易保护主义

technical barriers, technical barrier to trade (TBT) 技术壁垒是非关税壁垒的一类，它以技术为支撑条件，即商品进口国在实施贸易进口管制时，通过颁布法律、法令、条例、规定、建立技术标准、认证制度、卫生检验检疫制度、检验程序以及包装、规格和标签标准等，提高对进口产品的技术要求，增加进口难度，最终达到保障国家安全、保护消费者利益和保持国际收支平衡的目的

green trade barriers 绿色贸易壁垒，也称为环境贸易壁垒(Environmental Trade Barriers, ETBs)，它是国际贸易中的一种以保护有限资源、环境和人类健康为名，通过制定一系列苛刻的、高于国际公认或绝大多数国家不能接受的环保标准，限制或禁止外国商品的进口，从而达到贸易保护目的而设置的贸易壁垒

pesticide residues [ˈrɛzəˌdjʊ] n. 农药残留

system certification 体系认证

MRL, maximum residue limits of pesticide 农药最大残留限量。低于这个限量标准表示在长期食用情况下，对食用者无不良影响

organic tea 有机茶。该类茶叶的生产地没有任何污染，且是按照"有机农业"标准生产的鲜叶，没有非天然的添加剂，经过相关机构检验认证的茶叶

思考题

1. 将下面的句子译成英文。

（1）虽然全球茶叶消费量在增长，但增长率低于茶叶生产。国际茶叶委员会的统计显示，在2017年，全球茶叶消费量为557.1万吨，低于全球茶叶产量(581.2万吨)。

（2）世界茶叶出口占茶叶产量的比重从2008年的41.7%下降到2017年的30.8%，这进一步证实了世界茶叶生产大于消费的矛盾，茶叶生产和消费矛盾日益突出。

（3）优化出口茶类结构，发展绿茶、稳定红茶、提升特种茶(包含有机茶)出口，继续发挥我国作为世界第一大绿茶生产和出口国的优势。

2. 将下面的句子译成中文。

（1）Black tea consumptionby FAO is projected to grow at 2.5 percent annually to reach 4.17 million tons in 2027, reflecting the strong growth in consumption in producing countries, which should more than offset projected declines in traditional tea importing countries.

（2）Therefore, China should establish the assurance system of quality and safety, improve the proportion of organic tea gardens and ecological tea gardens, and sustainably develop physical control, biological control and other green prevention and control technologies to reduce the usage amount of pesticides and fertilizers.

(3) With the increasing health consciousness, we should optimize the structure of export tea to promote our advantage on high-quality green tea, special teas including organic tea, oolong, white, yellow and dark teas. Moreover, the exchange and integration of different tea cultures will further promote the development of tea trade in different countries.

参考文献

[1] China Tea Marketing Association. Analysis report on the situation of tea production and marketing in China in 2018[J]. Tea World, 2019(2): 10-15.

[2] LENG Y, SHANG H, SHI X, et al. A Brief Analysis of China's Tea Import and Export in 2018[J]. China Tea, 2019(4): 30-32.

[3] Committee on commodity problems, intergovernmental group on tea. CCP: TE 18/CRS1 Current market situation and medium term outlook[R]. Hangzhou, China: FAO, 2018.

[4] JIANG Z. Current situation and countermeasure of China's tea export trade[J]. Chinese Market, 2019(4): 64-67.

[5] 陈宗懋. 中国茶叶大辞典[M]. 北京: 中国轻工业出版社, 2008.

UNIT Ten　Tea Academic Paper Writing

一、科技论文结构

科技论文是科学技术人员或其他研究人员在科学实验(或试验)的基础上,对自然科学、工程技术科学等领域的现象(或问题)进行科学分析、综合的研究和阐述,并按照各个科技期刊的要求进行电子和书面的表达。茶学科技论文是世界各国茶学以及相关专业工作人员、研究人员、学者相互交流的主要手段。

1. 科技论文的特点

(1)学术性　学术性是科技论文的主要特征,它以学术成果为表述对象,以学术见解为论文核心,在科学实验(或试验)的前提下阐述学术成果和学术见解,揭示事物发展、变化的客观规律,探索科技领域中的客观真理,推动科学技术的发展。学术性是否强是衡量科技论文价值的标准。

(2)创新性　科技论文必须是作者本人研究的,并在科学理论、方法或实践上获得的新的进展或突破,应体现与前人不同的新思维、新方法、新成果,以提高国内外学术同行的引文率。

(3)科学性　论文的内容必须客观、真实,定性和定量准确,不允许丝毫虚假,要经得起他人的重复和实践检验;论文的表达形式也要具有科学性,论述应清楚明白,不能模棱两可,语言准确、规范。

2. 文本结构

茶学科技论文属于理工科论文,作为典型的应用科学文章或试验研究报告,其文本结构一般由标题、作者、摘要、关键词、前言、材料与方法、结果、讨论、结论等部分组成。

(1)标题页(Title page)

①标题:要求简明、准确地写出该课题研究的基本内容或研究揭示的创新点。优秀的茶学科技论文标题的特点是主题突出、明确,结构简洁、逻辑严谨,力求读者一见标题

就可以获得论文主旨和涉及的内容。有些杂志需要提供简短标题(running title)，在投稿时按照要求（一般少于6个单词）加在标题页。

②作者：包括所有作者的完整署名、所属单位及地址等。同时标记通信作者，并列出通讯作者的邮箱、联系电话、开放研究者与贡献者身份识别码(Open Researcher and Contributor Identifier, ORCID)等。针对共同贡献作者或共同通信作者，也在此处进行叙述。

（2）摘要页（Abstract page）

①摘要(Abstract)：是文章的缩影或主旨要点，需要作者概括地说明该研究的目的及重要性，并极其扼要地表述是以何种实验材料与方法得出的何种研究结论，突出论文的新见解和研究结果的意义。摘要根据投稿杂志要求，有字数限制（一般100~300字），因此需要以精炼的语言介绍文章的目的、论点、实验结果、结论、意义等。结构一般固定、内容简洁完整，应该让读者了解全文的梗概。

②关键词(Keywords)：被列入文献情报的检索系统，有利于全文的检索和交流。关键词一般为4~6个，列在摘要之后、正文之前，多用名词或名词词组形式，关键词之间一般用逗号或者分号进行分隔，最后一个关键词后面一般不加任何标点符号。

（3）引言和前言部分(Introduction)　引言和前言一般包括立项依据、研究现状、国内外研究进展、研究意义等方面。引言是科技学术论文正文的第一部分，它一般包括以下三个方面：介绍本研究的研究目的和研究意义；介绍相关的研究进展；进而引出本研究要解决的具体科学问题。也有文章在引言的最后会突出本研究或本文的研究内容，简单阐述一下这些研究内容或者结果带来的科学意义。引言在文中是独立而又完整的一部分，其篇幅可以是一至数段。

（4）材料和方法部分(Materials and methods)　材料和方法包括实验所用的植物材料、实验实施具体环境、实验设计或使用的方法、数据处理使用的软件等。根据杂志的要求和研究内容的不同，此部分有不同的写法，也可以有不同的标题或小标题。该部分内容在描述自己的方法时要尽可能的详细、要具备可重复性及操作性；使用别人的方法要引用；改进别人的方法时也需要清楚描述。

（5）结果和讨论部分(Results and discussion)　结果和讨论是文章的主要部分，是全文的重中之重。用图表准确列出实验中得出的数据；表述实验得出的最终结果并加以讨论。有的杂志要求分为结果(Results)和讨论(Discussion)两个部分，在文章讨论之后就可以结束论文了；也有的杂志要求将结果(Results)和讨论(Discussion)合并为一个部分进行阐述，那么在结果和讨论(Results and discussion)之后就需要有结论(Conclusions)部分。根据文章中涉及的不同的实验或者不同的结果，结果和讨论部分可以加入小标题分别加以阐述。结果往往是客观的阐述和介绍实验结果，讨论往往是客观的分析该实验结果以及形成该实验结果的科学机制和理论意义等。讨论是将实验研究中的感性认识提高到理性认识高度。其重点内容是对实验数据和现象进行科学分析，并对数据误差和影响实验结果的因素进行解释，探讨对实验材料及方法的改进。在讨论的撰写中，表述要全面、辩证、客观、切忌武断。对于学生的科技论文写作而言，结果与讨论部分，尤其是讨论部分是难点。

(6)结论部分(Conclusions) 科技论文一般要有结论,要求对文章做出清晰、明确的结论,写明本研究发现了哪些新的规律、发展了哪些学术理论、能解决什么现实问题。结论应全面、明确地突出本研究的科学发现,可以让读者再次进行思考。结论中可以包括一些对于该研究领域的建议、文章研究内容的应用前景等。然而要求文章的结果和讨论部分分开阐述的杂志,往往不需要一个独立的结论部分,但是在讨论的最后可以用一段话对全文进行总结,对该研究的科学意义进行进一步的阐述。

(7)致谢部分(Acknowledgements) 致谢一般写在正文之后,指为本研究提供帮助、指导、资助的单位和个人表示感谢。也有一些杂志需要以"Funding sources"形式列出获得资助的项目或基金,而不需要在致谢部分陈述。

(8)参考文献(Reference) 参考文献是指列出文中所有参考的资料,便于读者查看或检索与本文论述相关内容的直接资料。参考文献需要列出的内容包括作者、论文标题、期刊名、年份、卷、期、页等。参考文献的排序分为按照文献在正文出现的顺序或按照姓氏字母顺序排列。参考文献的编辑和管理软件有 Endnote、Reference manager 等,具体格式需要参考杂志的投稿指南。在我国国内发表的科技论文参考文献格式一般按 GB/T 7714—2015《信息与文献 参考文献著录规则》执行。

二、科技论文写作方法

科技论文的写作需要遵循一定的写作规范,每个投稿的杂志都有其投稿指南以供参考。以下将从写作前的准备、写作、投稿等三个方面进行介绍。

1. 写作前的准备

(1)确定论文的性质 根据目前国际 SCI 论文的内容性质,有下列分类:原创性研究(Original research article)、综述(Review)、约稿(Invited article, editorial, commentary)、短讯(Short communication)、给编辑的信(Letter to the editor)等。

(2)选定最佳目标投稿期刊 一般情况下,期刊的选择可考虑以下几个方面的因素:文章的论题,实验的结果及潜在意义是否适合期刊读者群,期刊已发表的文章与您论文内容关系的密切程度;期刊的投稿费用,影响因子及变化趋势,审稿时间;对于少数期刊,中国科研工作者还需考虑期刊的偏向性和成功率等方面。

(3)论文的著作权 关于论文的署名,我们需要考虑:期刊对作者的数量的限制;第一作者的确定,是否有并列第一作者;通讯作者的确定;中间作者的排列顺序(根据论文贡献大小);每个论文署名都必须经过作者同意等。

2. 英文论文的写作

(1)题目(Title) 科技论文题目应简明扼要地反映论文工作的主要内容,切忌笼统。论文题目应该是对研究对象精确、具体的描述,这种描述一般要在一定程度上体现研究结论。因此,论文题目不仅应告诉读者论文研究了什么问题,更要告诉读者这个研究得出的结论。

（2）摘要（Abstract） 科技论文的摘要，是对论文研究内容的高度概括，其他人会根据摘要检索一篇论文，因此摘要应包括对问题及研究目的的描述、对使用的方法和研究过程进行的简要介绍、对研究结论的简要概括等内容。通过阅读科技论文摘要，读者应该能够对论文的研究方法及结论有一个整体性的了解，因此摘要的写法应力求精确简明。大多数读者由文章的摘要来判断自己是否值得往下读全文。因此，摘要的描述要简练易懂，能提供主要信息并吸引大多数读者的注意，不宜太详细。我们从写作的顺序来简要介绍摘要的写法：首先，用一到两句话概括整个工作的内容；其次，用几句来对你的主要发现进行描述，介绍一些步骤过程和机理，重点提及与其他研究不同之处；最后，用一两句描述来总结此项的意义。

（3）前言（Introduction） 前言大致包含以下部分：问题的提出、选题背景及意义、文献综述、研究方法、论文结构安排等。也就是说讲清所研究的问题"是什么"；为什么选择这个题目来研究；对本研究主题范围内的文献进行详尽的综合述评；论文所使用的科学研究方法；本论文的写作结构安排等。各部分之间要存在有机联系，要有一定的逻辑顺序。

（4）材料及方法（Materials and methods） 这个部分的描述必要时要尽可能详细，便于别人可以重复你的实验，但也要避免过于详细。准确描述实验试剂的浓度、用量、实验温度、时间仪器型号，材料的来源。如果是动物实验，你必须提供实验许可以及遵守动物照管条例的证明；新发现的核酸或氨基酸序列需要在交稿前上传到 GenBank、DDBJ 等公开数据库。每个小部分以 1~2 句概括性的描述开头，用过去时态能包含结果，除非这个结果是方法本身实现的或在后续方法中要用到的。

（5）实验图表（Figures and tables） 通常情况下，读者单看论文的图表及图注、表注就可以知道这篇文章大体所做的工作。因此，图表要有很好的自明性。在作图颜色的选择上，需要注意的是颜色的选择和搭配。如果图片中有比例尺，要在图片标注中说明。

（6）结果（Result） 一些有经验的科研工作者有将结果部分作为最初开始动笔的地方。因为实验结果决定了哪些部分需要在方法中详细描述，哪些部分需要介绍或详细讨论。不是每个实验都必须要写在文章中，有些数据可以作为补充部分来呈现。在结果部分，应当简洁描述结果的本身。而对造成结果的原因或分析比较数据的差异等需要在讨论部分进行描述。

（7）讨论（Discussion） 在讨论的开始阶段，通常需要对论文所做的工作进行总结并阐述将来可能应用方面。讨论部分不是对结果的重复说明，而是要详细解释每一步实验是如何支持你的主题论点。你也可以加上一些文献材料来再次支持你的观点，并强调你的发现对这个领域的研究有较为重要的意义。在讨论的最后，通常需要指出此项工作或在该领域还待大家探讨的问题。

3. 英文论文投稿

（1）同行审阅（Peer review） 在投稿的时候可以自己推荐或者排除审稿人。但是不能推荐同一单位、朋友、共同第一作者（三年内）、合作者（提供给你实验材料等）等。对于排除审稿人，合适数量通常是 2 个左右，并分别简要写明排除的原因，如财务支持上的竞争、观点上的偏见等。对于《美国科学引文索引》(SCI)收录期刊的编辑，他们通常在选择审稿人的时候也是通过 PubMed 的搜索，选择一些有审稿经验的、比较负责的、并且比较公正的同一领域研究学者。

（2）修改(Revision) 通常情况下，如果文章不被拒收的话，期刊主编都会提出不同程度修改意见。如果编辑回复并特意列举了一些需要修改的地方，这就意味着这些是修改的重点。同时，也必须照顾到每个审稿人的意见。充分全面的回复修改意见是文章能被接收的前提，不要自己选择性地回复，这也是你对他们意见的重视程度的表现。除了这些意见，你必须同时在文章中的相应位置明确标注哪些是已经被修改过的。如果你确实不能满足审稿人的建议，你必须提供合理的原因来解释为什么这些修改在目前的情况下能以实现。虽然这些理由不一定总是有用，但编辑会慎重考虑你的解释。

（3）接收(Acceptance) 文章被接收后，就会转到出版商那里，出版编辑会在 1~2 天或 2~3 周或更长时间内对稿件进行排版和相应的语法修改，这个工作是在内部排版部门进行的，无法跟进，一般也没有办法催促。准备就绪后，出版方会将排版好的稿件发给通讯作者进行文章校对"Article Proof"。一般清样的校对时间比较短（一些杂志要求 24~48 小时内返回），需要尽快回复。校对时最重要的是确认作者署名、单位和邮件地址的准确性，因为一旦校对完成就无法对文章再做任何修改。随着信息化的发展，目前一些期刊都能够进行网上在线校对，这无疑节省了作者时间，值得赞赏。这时请务必认真检查，因为这是最后一次确认作者信息及文章内容的机会。此外，校对后作者还应该留意单行本订阅、版面费和彩图费支付等问题。校稿完成返给出版商之后，文章就会进入"Article in Press"阶段，这时你的文章可以在网上进行查阅，当然文章正式出版需要看杂志的安排，一般要几个月或者更长的时间。

三、科研诚信

科研诚信(Scientific integrity)是科技创新的基石。近年来，我国科研诚信建设在工作机制、制度规范、教育引导、监督惩戒等方面取得了显著成效，但整体上仍存在短板和薄弱环节，违背科研诚信要求的行为时有发生。为此，中共中央办公厅、国务院办公厅在 2018 年 5 月印发了《关于进一步加强科研诚信建设的若干意见》，提出加强顶层设计、构建符合科研规律、适应建设世界科技强国要求的科研诚信体系；加强科研活动全流程诚信管理，严肃查处严重违背科研诚信要求的行为，对严重违背科研诚信要求的行为依法依规终身追责。为此，国家新闻出版署发布并于 2019 年 7 月 1 日实施的 CY/T 174—2019《学术出版规范—期刊学术不端行为界定》中对论文作者、审稿专家、编辑者所可能涉及的学术不端行为进行了界定。如剽窃(Plagiarism)是指采用不当手段，窃取他人的观点、数据、图像、研究方法、文字表述等并以自己名义发表的行为；伪造(Fabrication)指编造或虚构数据、事实的行为；篡改(Falsification)指故意修改数据和事实使其失去真实性的行为；不当署名(Inappropriate authorship)指与对论文实际贡献不符的署名或作者排序行为；一稿多投（Duplicate submission；Multiple submission）指将同一篇论文或只有微小差别的多篇论文投给两个及以上期刊，或者在约定期限内再转投其他期刊的行为；重复发表（Overlapping publication）是指在未说明的情况下重复发表自己（或自己作为作者之一）已经发表文献中内容的行为。此标准对学术不端行为进行具体的界定，适用于学术期刊论文出版过程中各类学术不端行为（Academic misconduct）的判断和处理。其他学术出版物可参照使用。

治理包括论文代写在内的弄虚作假、学术不端行为，加强大学生的学术诚信建设、净化高校学术风气，不仅事关学风，更事关世风，是一项必须重视的系统工程。2018年7月，教育部印发《教育部办公厅关于严厉查处高等学校学位论文买卖、代写行为的通知》（教督厅函［2018］6号），要求对参与购买、代写学位论文的学生给予开除学籍处分，已获得学历证书、学位证书的依法予以撤销，被撤销的学历证书、学位证书已注册的应予以注销并报教育行政部门宣布无效。

John D'Angelo博士指出的科学研究中的道德失范问题如下。

Intentional negligence in the acknowledgment of previous work (including work you did). In this instance, a researcher is intentionally not mentioning work that has already been done in the field. The same violation is occurring when an author only acknowledges work inferior to his or her own. It is the researcher's absolute responsibility to make at the very least a good faith effort to search all of the relevant literature before proceeding with a publication or grant.

Deliberate fabrication of data you have collected. In this ethical violation, researchers quite literally make up data, claiming experiments were carried out when they were not or significantly altering the results they do obtain so that they fit the pre-experiment hypothesis or previous studies. It also has the unfortunate consequence of leading other researchers down an incorrect and potentially impossible path trying to repeat and/or use the fabricated results. This then has many negative effects on other researchers' careers, resulting in wasted time and research funds.

Deliberate omission of known data that does not agree with the hypothesis. If any particular results are left out of a publication just because they do not agree with pre-experiment hypotheses, an ethical violation has certainly been committed.

Passing another researcher's data as one's own. This falls into the category of plagiarism. One variety is where a researcher reads a paper or grant, or attends a seminar, and then attempts to publish the data as is, as his own work.

Publication of results without the consent of all of the researchers. One frequently overlooked ethical violation is that all authors must approve the publication. When it is overlooked, it is often an honest error and can be resolved peacefully. There are cases, however, that can be imagined where one of the authors does not agree with one of the chief conclusions the other authors are putting forth. In a case such as this, a resolution must be reached before the publication can ethically proceed. There are also cases where one of the authors may not even be aware of the paper; this is not unthinkable, since sometimes parts of an author's work may get published years after leaving a lab. In gross misconduct cases, an author who may have nothing to do with an experiment is added to a paper to give it more "weight". The reason why it increases the chances of the paper getting published is that many reviewers and editors may be more hesitant to make negative comments on an article co-authored by a preeminent scientist.

Failure to acknowledge all of the researchers who performed the work. This ethical violation is subject to a great deal of gray area, perhaps even the largest amount of gray area. With the high monetary stakes involved, especially with patents, this is an enormously important issue. On the other hand, it is also important to not be too generous—adding undeserving authors has the effect of diminishing each author's apparent contribution.

Conflict of interest. When authors submit a manuscript of any type or format they are responsible for disclosing all financial and personal relationships that might bias or be seen to bias their work.

Repeated publication of too-similar results or reviews. Oftentimes, researchers try to boost their publication records by publishing multiple times on the same work or too-closely related work. This is not unlike plagiarism and in fact is a form of plagiarism called "selfplagiarism".

Breach of confidentiality. Breach of confidentiality may be when a reviewer does not keep information for herself but passes it on to a colleague who is competing in the same area as the author with the work under review.

Misrepresenting others' previous work. One example of this violation is to present your own (and aberrant) conclusions of someone else's work as their conclusions.

We will advocate academic integrity, encourage independent thinking, ensure academic freedom, and foster a scientific spirit(倡导学术诚信，鼓励独立思考，保障学术自由，弘扬科学精神).

参考文献

[1] 屠康，贺稚非. 食品专业英语[M]. 2版. 北京:中国农业出版社，2015:1-22.

[2] ALBERT T, WAGER E. How to handle authorship disputes: a guide for new researchers[D/OL]. The COPE Report 2003. https://publicationethics.org/files/2003pdf12_0.pdf.

[3] D'ANGELO J. Ethics in science—Ethical misconduct in scientific research[M]. Boca Raton CRC Press, 2012.

[4] GB/T 7714—2015. Information and documentation—Rules for bibliographic references and citations to information resources(信息与文献　参考文献著录规则)[S].

[5] CY/T 174—2019. Academic publishing specification—Definition of academic misconduct for journals(学术出版规范　期刊学术不端行为界定)[S].

Appendix

Appendix A: Websites and Associations (Societies) on Tea

网址	英文名称	中文名称
www.chinatss.cn	China Tea Science Society	中国茶叶学会
www.ctma.com.cn	China Tea Marketing Association	中国茶叶流通协会
www.tc339.com	National Technical Committee 339 on Tea of Standardization Administration of China	全国茶叶标准化技术委员会
www.iso.org	International Organization for Standardization	国际标准化组织
www.fao.org	Food and Agriculture organization of the United Nations	联合国粮农组织
www.inttea.com	International Tea Committee	国际茶叶委员会
www.eatta.com	East African Tea Trade Association	东非茶叶贸易协会
www.tea.co.uk	UK Tea&Infusions Association	英国茶叶与饮料协会
www.irishteatrade.ie	Irish Tea Trade Association	爱尔兰茶叶贸易协会
www.teausa.com	Tea Association of the U.S.A	美国茶叶协会
www.tea.ca	Tea & Herbal Association of Canada	加拿大茶叶协会
www.teaboard.or.ke	Tea Board of Kenya	肯尼亚茶叶委员会

续表

网址	英文名称	中文名称
www.ktdateas.com	Kenya Tea Development Agency	肯尼亚茶叶产业发展局
www.teaboard.gov.in	Tea Board of India	印度茶叶局
www.indiatea.org	Indian Tea Association	印度茶叶协会
www.upasi.org	United Planters' Association of Southern India	南印度种植者协会
www.tocklai.org	Tea Research Association Tocklai	托克莱茶叶研究协会
www.pureceylontea.com	Sri Lanka Tea Board	斯里兰卡茶叶局
www.srilankateaboard.lk	Sri Lanka Tea Board	斯里兰卡茶叶局
www.rusteacoffee.ru	Russian Association of Tea & Coffee Producers	俄罗斯茶叶与咖啡生产者协会
www.o-cha.net	World Green Tea Association	世界绿茶协会
www.tea-a.gr.jp	Japan Black Tea Association	日本红茶协会
www.indonesiateaboard.org	Indonesia Tea Board	印度尼西亚茶叶委员会
www.vitas.org.vn	Vietnam Tea Association	越南茶叶协会
www.pakistanteaassociation.com	Pakistan Tea Association	巴基斯坦茶叶协会
http://teaboard.go.tz	Tea Board of Tanzania	坦桑尼亚茶叶局
www.teaboard.gov.bd	Bangladesh Tea Board	孟加拉国茶叶委员会
http://teacoffee.gov.np	National Tea and Coffee Development Board (Government of Nepal)	尼泊尔国家茶和咖啡发展委员会
http://ugatea.com	Uganda Tea Development Agency	乌干达茶叶开发署
www.rainforest-alliance.org	Rainforest Alliance	雨林联盟
www.ethicalteapartnership.org	Ethical Tea Partnership	茶叶道德联盟
www.china-teaexpo.com	China International Tea Expo	中国国际茶叶博览会(杭州)
www.worldteaexpo.com	World Tea Expo	世界茶叶博览会(美国)

Appendix B: Tea Sensory Terminology

Table B.1 茶叶外形与叶底的概念性用语

名称	释义	用法说明	名称	释义	用法说明
毫 Pekoe/Tippy	芽上细长而尖的毛	外形	片/张 Leaf	完整或不完整的叶片	外形/叶底
芽 Bud	处于幼态未伸展的枝芽	外形/叶底	末 Dust	细砂粒状或粉末状茶	外形
茎 Stem	未木质化的枝条	外形/叶底	条 Strip	条状干茶	外形
梗 Stalk	木质化的枝条	外形/叶底	块 Lump	稍大的块状干茶	外形
筋 Fibre	脱去叶肉的叶柄、叶脉	外形/叶底	头 Knob	圆块状干茶	外形
皮 Epidermis	木质化枝条的表皮层	外形/叶底	斑/点 Spot	由于工艺原因在干茶或叶底形成的斑点	外形/叶底

Table B.2 茶叶外形与叶底的描述性用语

分类	名称	释义	用法说明
重量 weight	轻/重 Light/Heavy	干茶密度小/大	外形
紧结度 Tightness	紧/结、实/松 Tight/Sturdy/Loose	干茶形状紧/结实/不紧	外形
大小 Size	硕、壮/肥/瘦 Burly/Thick/Thin 大/小 Big/Small 粗/细 Coarse/Slender	硕、壮:形容中、大叶种较大的干茶外形 肥:嫩度好,芽叶柔软厚实 瘦:嫩度差,芽叶单薄少肉 大/小:比正常规格大/小 粗/细:嫩度差/好,形状粗大/细小	外形/叶底
形状 Shape	尖、削/挺、直/盘、钩、卷、弯、曲、扭 Sharp/Straight/Curly	扁茶芽尖如剑峰/茶条不弯曲,形状直/茶条弯曲(从盘到扭程度逐渐减弱)	外形(扁形)/扁形、条形/条形
	圆、浑/扁、平 Round/Flat	条索或颗粒呈圆形/茶条外形扁	外形(颗粒形、条形)/扁形、条形
	宽/窄、狭 Wide/Narrow	扁形茶长宽比不当,宽度过大/小	外形(扁形)
	长/短 Long/Short	茶条过长/短	外形(扁形、条形)
	折、皱/展 Crinkle/unfold	干茶或叶底起褶皱/开展	外形/叶底

续表

分类	名称	释义	用法说明
嫩度 Tenderness	嫩/老 Tender/Over mature	嫩度好/差	外形
完整度 Integrity	整/碎 Unbroken/Broken	干茶完整/断碎	外形/叶底
均匀度 evenness	匀/杂 uniform/uneven	干茶匀齐/不匀齐、杂乱	外形/叶底
净度 Purity	净 Clean	干茶洁净	外形/叶底
光滑度 Smoothness	光、滑/毛 Smooth/Rough	干茶或叶底表面光滑/干茶表面毛糙	外形、叶底/叶底
软硬度 Hardness	柔、软/硬 Soft/Hard	叶底柔软/硬	叶底
厚薄度 Thickness	厚/薄 Thick/Thin	叶底叶张厚度好/差	叶底

Table B.3　　茶叶颜色相关术语

名称	常用组合	用法说明
白 White	银白、灰白	外形/叶底
黄 Yellow	绿黄、深黄、杏黄、浅黄、橙黄、姜黄、灰黄、褐黄、棕黄、金黄、清黄、浅杏黄、青黄、蜜黄、绿金黄	外形/汤色/叶底
绿 Green	浅绿、靛青(靛蓝)、黄青、深绿、银绿、绿艳、嫩绿、碧绿、杏绿、黄绿、翠绿、灰绿、墨绿(乌绿、苍绿)、砂绿、青绿、蜜绿	外形/汤色/叶底
红 Red	橙红、深红、栗红、褐红、微红、浅红、棕红、紫红、糟红、粉红、红、红艳	外形/汤色/叶底
紫 Purple	紫、微紫、红紫、青紫	外形/汤色/叶底
褐 Brown	红褐、绿褐、青褐、黄褐、灰褐、棕褐、栗褐、黑褐、乌褐	外形/叶底
乌(黑) Black	乌、铁黑、褐黑	外形/叶底

Table B.4　　茶叶色觉术语

分类	名称	释义	用法说明
光泽度 Glossiness	油、润/枯 Glossy/Sere	干茶光泽好/缺乏光泽	外形
清澈度 Clarity	清、澈/混、浊 Clarity/Turbid	茶汤透明、光亮/有悬浮物，透明度差	汤色
明亮度 Brihtness	明、亮/暗 Bright/Dark	茶汤或叶底反光强/弱	汤色/叶底
鲜艳度 Vividness	鲜、艳 Vivid	艳丽的色系	汤色/叶底
深浅度 Depth of colour	深/浅 Deep/Light	色泽深/淡	外形、汤色/汤色
均匀度 Evenness	花 Mixed	叶色不一	外形/叶底

Table B.5 茶叶味觉术语

分类	名称	释义
浓度味型 Intensity	浓/强 Strong	内含物丰富，收敛性强
	醇 Mellow	浓淡适中，口感柔和
	和 Light	入口稍淡，无刺激性
	淡 Weak/Pale	入口寡淡无味
特征味型 Basic tastes	甘/甜 Sweet	有甜味
	鲜 Umami	有氨基酸的味道
	苦 Bitter	有苦味
	酸 Sour	有酸味
感觉味型 Mouthfeel	厚 Body/Thick	茶汤内含物丰富，有粘稠感
	薄 Thin	茶汤入口稀薄
	滑/润 Smooth	入口和吞咽顺滑、无粗糙感
	涩 Astringent	厚舌阻滞的感觉
	糙 Harsh	舌面粗糙的感觉

Table B.6 茶叶香气术语

分类	名称	释义
香气类型 Types of aroma	鲜(香) Fresh	鲜爽感
	嫩(香) Tender	嫩茶所特有的愉悦细腻的香气
	栗(香) Chestnut	类似熟栗子的香气
	清(香) Faint scent	类似绿色植物清新纯净的香气
	甜(香) Sweet	甜的香气
	花(香) Flowery	类似鲜花的香气
	馥/郁/幽(香) Fragrant	花香的三种不同类型
	蜜(香) Honey	蜂蜜的香气
	果(香) Fruity	浓郁的果实熟透的香气
	焦糖(香) Caramel	类似蔗糖或红糖的香气
	木(香) Woody	梗叶木质化，带有纤维气味
	烟(香) Wood smoky	松枝或柴枝熏烤后产生的烟香
	烟(气) Burnt smoky	茶叶烧焦后产生的烟气
	炭(香) Baked carbon	炭火烘焙后产生的香气
	火(香) Fired	干燥充分的新茶具有的香气
	火(气) Over fired	似锅巴香，干燥过程温度高或时间长产生的香气

续表

分类	名称	释义
香气类型 Types of aroma	焦(气)Burnt	严重的焦烟气
	陈(香)Aging	茶质好、保存得当，陈化后具有的愉悦香气
	陈(气)Stale	存放中失去新茶特征产生的不愉快气味
	闷(气)Stuffy	沉闷不爽的气味
	晒(气)Sun shined	日光照射后产生的日光味
	生(气)Over green	浓烈的青草或青叶的气味
	青(气)Green	带有青草或青叶的气味
	粗(气)Coarse	粗老叶的气味
	酸(气)Acid	似醋般酸的气味
	馊(气)Sour	似泔水般馊的气味
	堆(气)Off odour of pile	黑茶渥堆发酵产生的气味
	仓(气)Off odour of storage	黑茶储存不当产生的气味
	霉(气)Mouldy	茶叶存放中水分过高微生物生长分泌的具有霉感的气味
	油(气)Oily	茶叶加工过程中加入炒茶油，干茶存放后产生的气味
	粟(香)Parched rice	中等火温长时间烘焙产生的如粟米的香气
	奶(香)Milk	似乳制品的香气
	酵(气)Over fermentated	不应发酵的茶类产生的发酵气味
	辛(香)Pungent	青辛气味
	毫(香)Pekoe aroma	茸毫含量多的茶叶特有的香气
香气强度、格调与持久度 Aroma intensity, style and durability	浓/强/高 Strong	香气的浓度高
	淡/弱/低 Weak	香气的浓度低
	平 Normal	香气特征不突出，平淡
	飘/薄/浮 Light	香气浮于表面，一嗅而逝
	锐 Sharp	香气尖锐
	钝 Dull	香气圆钝
	纯/正 Pure	香气纯正
	爽 Brisk	香气爽快，令人愉悦
	浊 Unbirsk	香气混浊
	长 Lasting	香气持久
	短 Unabiding	香气不持久

Table B.7 茶叶感官审评程度描述术语

名称	释义	用法说明
高/强 high/Strong	程度高	形容香气、滋味
低/弱 Low/Weak	程度低	形容香气
深 Deep	颜色深	形容色泽
浅 Light	颜色浅	形容色泽
较 Comparatively	比较用词，较高程度的具有某特征	褒贬通用
略 Comparatively	比较用词，较高程度的具有某特征	贬义专用
稍 A little	比较用词，较低程度的具有某特征	褒贬通用
微 Slightly	比较用词，极低程度的具有某特征	褒贬通用
尚 Barely	比较用词，极低程度勉强具备某特征	褒义专用
欠 Not enough	比较用词，缺乏某特征	褒义专用
有 With the character of	具有某特征	褒贬通用
多 Highly with the character of	较多的具有某特征	褒贬通用
显 Obviously with the character of	明显的具有某特征	贬义专用

引自：张颖彬，刘栩，鲁成银．中国茶叶感官审评术语元语素研究与风味轮构建［J］．茶叶科学，2019，39（4）：474-483．

Appendix C： Introduction to State Key Laboratory of Tea Plant Biology and Utilization

　　Tea science in Anhui Agricultural University is a leading teaching and research program in China. The tea science program was established in 1939 at Fudan University, and moved to Anhui Agricultural University in 1952. The founders were Professors Zenong Wang and Chuan Chen who were two pioneers of tea sciences in China. In 2003, it was selected by Ministry of Science and Technology of China to construct the province and ministry co-founded state key laboratory. For 12-year continuous improvement, it was officially approved as "State Key Laboratory of Tea Plant Biology and Utilization" in January 2015, and was supported by Ministry of Science and Technology of China and Government of Anhui province.

　　The key laboratory has been a leading laboratory in the fields of tea science research and technology innovation for tea industry. The key laboratory has 65 regular faculties, including 27 full professors, 20 doctoral tutors; and most of them had experiences of studying and researching in developed countries.

　　The key laboratory has 4 research interests as following: Physiological Ecology and

Germplasm Innovation of Tea Plant, Metabolism and Quality Chemistry of Tea, Health Beneficial Effect and Mechanism of Tea, Tea Quality & Safety and Tea Processing & Utilization, which are all focused on basic scientific researches and its applications.

The key laboratory has 12,000m^2 laboratory space, and 96 million RMB research facilities. The key laboratory has established 8 facilities centers, which are Center for Genomics, Center for Metabolomics, Center for Proteomics, Center for Cell Biology, Center for Physiology and Nutrition, Center for laboratory Animals, Center for Tea Quality Control, and Center Tea Processing.

In the past five years, the tea laboratory has funded by more than 90 research project, including national key technology R&D program of China, key project of 973, international cooperation foundation, national natural science foundation of China, and revitalization project for Anhui tea industry.

Also, tea laboratory has earned numerous award, including: Science and Technology Innovation Team of Anhui Province, 2006; 115 Industry Innovation Team of Anhui Province, 2008; Changjiang Scholars and Innovative-Research-Team Ministry of Education of China, 2011; Top 10 Outstanding Industries Team of Anhui Province, 2013 and National Outstanding of Professional Team, 2014.

Our state key laboratory researches focus on the innovation of tea science and technology, meet the national mid-long term developmentneeds for tea industry. The goals of our laboratory are try to provide new solutions for scientific and technical issues of tea science, pursue innovation researches and cultivate high-level talent team for the development of tea industry in China.

Our laboratory is opening positions for high-level talents world wide, with background of Genetics and Breeding of Tea Plant, Genomics of Plant, Secondary Metabolites of Plant, Natural Products Chemistry of Plant, Tea Processing and Quality Control, and Prevention of Metabolic Syndrome. The research interests will cover Germplasm and Utilization of Tea Resources, Genomics of Tea Plant, Physiological Ecology of Tea Plant, Processing Chemistry and Quality control of Tea and Tea for Health.

Address: 130 West Changjiang Road, Biological S&T Building, Anhui Agricultural University, Hefei 230306, Anhui Province, China

Tel: +86-551-65786765, 65786401

Fax: +86-551-65786765

Website: http://tealab.ahau.edu.cn